Research Methodology and Strategy

Theory and Practice

Patrick X.W. Zou, PhD
Chang'an Scholar Distinguished Professor
Chang'an University, China

Xiaoxiao Xu, PhD
Associate Professor
Nanjing Forestry University, China

WILEY Blackwell

Registered Office(s)
John Wiley & Sons, Inc., 111 River Street, Hoboken, NJ 07030, USA
John Wiley & Sons Ltd, The Atrium, Southern Gate, Chichester, West Sussex, PO19 8SQ, UK

For details of our global editorial offices, customer services, and more information about Wiley products visit us at www.wiley.com.

Wiley also publishes its books in a variety of electronic formats and by print-on-demand. Some content that appears in standard print versions of this book may not be available in other formats.

A catalogue record for this book is available from the Library of Congress

Hardback ISBN: 9781394190225; epub ISBN: 9781394190232; epdf ISBN: 9781394190249; obook ISBN: 9781394190256

Cover Images: Courtesy of Patrick X.W. Zou and Xiaoxiao Xu, Marko Geber/Getty Images, David Buffington/Getty Images, artpartner-images/Getty Images
Cover Design: Wiley

Set in 9.5/12.5pt STIXTwoText by Integra Software Services Pvt. Ltd., Pondicherry, India

Our deepest gratitude to our families for their enduring and selfless love and support: Dr Hongyu Li, Joanne Zou, Daniel Zou, Yu Gao, Zimo Xu

Contents

Foreword by Andrew Dainty *vii*
Foreword by Chimay J. Anumba *ix*
Foreword by Lieyun Ding *xi*
Foreword by Dongping Fang *xiii*
Acknowledgements *xv*

1 **Fundamentals of Research** *1*

2 **Qualitative Research** *19*

3 **Quantitative Research** *37*

4 **Mixed Methods Research** *85*

5 **Case Study Research** *97*

6 **Technology-Enabled Experimental Research** *117*

7 **Data-Driven Research** *129*

8 **The Fifth Research Paradigm: Hybrid Natural-Social Sciences Methods Research** *151*

9 **Journal Article Writing and Publishing** *161*

10 **Thesis Writing** *187*

11 **Research–Practice Nexus and Knowledge Coproduction** *199*

12 **Managing the Researching–Writing–Publishing Journey** *211*

13 **Improving Impact and Citation of Research Outcomes** *225*

14 **Concluding Remarks and the Ways Forward** *235*

Index *237*

Foreword

Professor Andrew Dainty

Developing a research design and methodology for any study is one of the most fundamental decisions any researcher can make, and one which challenges many, particularly at the outset of their careers. Questions such as how to identify appropriate research problems and define research aims, what are the differences between various research methodologies and paradigms, what factors inform the choice of methods and how to publish impactful research outcomes are common questions that concern most researchers, and are those that this book sets out to answer.

Researcher Methodology and Strategy: Theory and Practice is different from many other books as it contains research methodology and strategy in one single volume. This book comprehensively describes research methodologies and approaches including qualitative research, quantitative research, and mixed methods approaches. It also discusses new emerging research methods such as technology-enabled experimental research methods and data-driven research methods. In addition, it explains emerging ideas such as the fifth research paradigm: hybrid natural-social sciences methods research, an exciting development in methodological thinking.

This book also provides comprehensive strategies for implementing research, including journal article writing and publishing, thesis writing, operating at the research–practice nexus and knowledge coproduction. It therefore provides practical guidelines for improving research theoretical and its impact.

I have known Professor Patrick X.W. Zou for many years, having collaborated in research that has achieved significant outcomes, including award-winning papers. He has developed a book suitable for students to improve the depth and breadth of research knowledge and skills, for researchers to improve research thinking and strategies, and for practitioners to improve knowledge coproduction and practical performance and I commend it to you.

Good luck with your research!

Andy Dainty

Professor Andrew Dainty, PhD
Pro-Vice-Chancellor
Manchester Metropolitan University, UK

Foreword

Professor Chimay J. Anumba

The need for appropriate use of research methods has been growing in importance over the years. In fact, many universities now have compulsory research methods courses that all postgraduate students have to take as part of their advanced studies and research. These students rely on a wide range of books to find the most relevant for their research.

Professor Patrick Zou and his colleague, Xiaoxiao Xu, have written a comprehensive book on *Research Methodology and Strategy: Theory and Practice*, as their contribution to enhancing understanding of research methods and strategies. It is different from other similar texts in the way that it goes from basic definitions of research to traditional and modern research approaches and then to thesis and journal article writing, and the importance of impact and citation of research outcomes. It also addresses the gap between research and practice, and includes review questions and exercises at the end of each chapter.

The book provides a very good introduction to the fundamentals of research and the key topics covered within the book. It fully describes qualitative, quantitative, and hybrid research methods, and provides guidance on their use and how to avoid the main criticisms associated with them. It also provides appropriate examples, as needed, to illustrate and/or emphasize the key points. Case studies, which are now very widely used are also well covered. A new topic that is not covered by many current books on this subject is 'Technology-Enabled Experimental Methods', which discusses electroencephalography (EEG) and eye tracking, and the context for their experimental use. The book also discusses the importance of data, data quality, and data analytics – all of which are now of considerable interest to the research community. The authors also introduce the Fifth Research Paradigm, which involves hybrid natural and social sciences research.

An important feature of this book is the coverage of research writing – both thesis and journal papers – and some of the issues involved in getting published. This is a welcome addition and will be of tremendous benefit to young researchers. The discussion of the gap between research and practice is also a novel feature that will enable researchers and practitioners to understand how best to bridge the gap, and enhance our collective knowledge.

Overall, this is a much-needed and timely book that will be invaluable to both research-ers and practitioners. It would be a great text for the numerous research methods courses that I mentioned at the outset. I commend it to all those interested in improving their research and research outcomes, and making an impact.

Professor Chimay J. Anumba, BSc, PhD, DSc, Dr. h.c.,
FREng, CEng, FICE, FIStructE, FASCE, NAC
Dean, College of Design, Construction, and Planning
University of Florida, USA

Foreword

Professor Lieyun Ding

The world we live in is complex, as are the natural and social problems we face. This is coupled with rapid development and constant change, and the increasing application of emerging information and communication technologies and data science. Traditional research methods have become insufficient in decision making, and in identifying, analysing, and solving complex problems. A series of questions continue to beset researchers: What methodologies should I apply? What methods do I use? What strategies should I take?

This book, *Research Methodology and Strategy: Theory and Practice*, focuses on not only research methodologies but also research strategies. It includes key aspects of scientific research and provides contemporary research methods and strategies to improve the efficiency, quality, and impact. There are several unique features: the development of the fifth research paradigm framework, technology-enabled research methods, data-driven research methods, and emphasis on the research–practice nexus and knowledge coproduction as well as a longitudinal perspective of research from conceptualization to long-term impact.

This book is easy to understand, learn from, and then apply the theories and techniques and practice them in different research contexts. It provides numerous contemporary theories and practical guides and examples. It will be of great value for higher degree research students to learn research methods and thesis writing, for researchers to improve research quality, outcomes, and impact, and for practitioners to improve the theoretical underpinning of their practice.

This book is a condensed reflection of Professor Patrick X.W. Zou's over 20 years of rich experience and knowledge in researching, teaching, and supervising from the frontline of research and practice.

I highly recommend this book to researchers, students and practitioners.

Professor Lieyun Ding, PhD
Academician, Chinese Academy of Engineering
Former President, Huazhong University of Science and Technology, China

Foreword

Professor Dongping Fang

Research mainly aims at fulfilling humankind's curiosity, solving problems, and developing strategies for the advancement of society and improvement of the natural environment. How to define research problems and aims? How to select research methods? How to publish research outcomes and improve research impact? These are common but key issues that require answers when undertaking a research project.

This book provides the answers to these questions. Not only so, the book also helps readers improve the depth and breadth of their research thinking and capability. This book explicitly discusses the fundamental concepts, theories, and techniques as well as practical processes, and uses numerous examples to explain different research methods in a simple and straightforward manner. Readers will also gain an in-depth understanding of research process, from problem formulation to outcomes publication and improving research impact.

Out of the many unique features, I am particularly impressed with the technologies-enabled and data-driven research methods as well as research strategies that cover thesis writing, journal article writing, submission, responses to reviewer comments, and monitoring publication impact. I am also impressed with the chapters discussing strategies on improving the research–practice nexus and knowledge coproduction. There are many worked examples and diagrams to explain the complex concepts and techniques.

I have known the lead author Professor Patrick X.W. Zou for more than 20 years, and I have also had the honour of collaborating with him. This book draws from Professor Zou's rich experience and knowledge of research, practice, and student supervision, in methodology and strategy.

I strongly recommend this easy-to-read book to anyone who wishes to learn research methodology and strategy, and improve research productivity, quality, and impact.

Professor Dongping Fang, PhD
Dean, School of Civil Engineering
Tsinghua University, China

Acknowledgements

We would like to thank the many colleagues, friends, and collaborators with whom we have studied and worked in various universities and industries. In particular, we would like to express our sincere appreciation to our colleagues at Chang'an University and Nanjing Forestry University. Special thanks to Professor Andrew Dainty of Manchester Metropolitan University for many valued discussions and collaborations on research methodologies and broader topics. Sincere thanks also go to Professor Lieyun Ding of Huazhong University of Science and Technology and Professor Dongping Fang of Tsinghua University for their continued support and collaboration. We would like to convey our heartfelt gratitude to Professor Chimay Anumba of the University of Florida, a Fellow of the Royal Academy of Engineering, for his continued support.

We would like to thank Wiley Blackwell for its support in publishing this book, in particular publisher for the Built Environment, Dr Paul Sayer, for his continued support and advice. We also sincerely thank the book proposal reviewers for their excellent comments to improve the quality of this book. We would like to express our gratitude to Elsevier and Springer Nature for granting us the copyright permissions to use their publications.

We are very grateful to Ms Joanne E. Zou and Dr Hongyu Li who have helped proofread the book and provided many valuable comments and discussions.

We gratefully acknowledge the support of the National Natural Science Foundation of China (Grant No. 72101118) and the Fundamental Research Funds for the Central Universities of China (Grant No. 300102231301).

Last but not least, we express our deepest gratitude to our families for their enduring and selfless love and support.

Professor Patrick X.W. Zou, *PhD*
Associate Professor Xiaoxiao Xu, *PhD*
Xi'an, China

1

Fundamentals of Research

1.1 Introduction

The advancement of human society is dependent on creating and applying new theory and new knowledge. This is achieved through research, which in turn requires the application of methodology and strategy.

Research is defined as the detailed study of a subject, especially in order to discover information or reach an understanding (Cambridge Dictionary), or an endeavour to discover new or collate old facts by the scientific study of a subject or by a course of critical investigation (The Oxford Encyclopaedic English Dictionary). It is a systematic process of collecting, analysing, and interpreting data to increase understanding of a phenomenon. Specifically, it is the systematic, controlled, empirical, and critical investigation of hypothetical propositions about presumed relations among natural or social phenomena. Research is a combination of experience and reasoning and is regarded as an approach to the discovery of truth, where experience leads to knowledge and understanding through day-to-day living and reasoning is a method of coming to a conclusion by the use of logical argument.

The main objectives of engaging in research are to develop new theory or support existing theory, and to create new knowledge. Theory is a statement about a phenomenon or a set of statements describing the interrelationships of the elements within a phenomenon, while knowledge is the understanding of or information about a subject that researchers get by experience or study.

Quality research can bring a range of social, cultural, and economic benefits locally and globally, leading to social development and productivity and economic improvement. At an individual personal level, successfully undertaking research could be a major development and achievement. It is an opportunity to master different research methods and methodologies, develop decision-making and problem-solving capabilities, and develop personal attitude, skills, and knowledge (ASK), which leads to a more professional approach and career opportunities.

This chapter introduces the characteristics and cornerstones of research, philosophical assumptions, general research process, different research paradigms and methods, theory development methods, research ethics, and an overview of the book's contents and its unique features.

Research Methodology and Strategy: Theory and Practice, First Edition. Patrick X.W. Zou and Xiaoxiao Xu.
© 2023 John Wiley & Sons Ltd. Published 2023 by John Wiley & Sons Ltd.

1.2 Characteristics of Research and the General Research Process

Research should have the following characteristics, as stated by Leedy and Ormrod (2013):

1) Originates with a research question or a problem; in other words, it is guided by a specific research problem, question, or hypothesis.
2) Divides the principal problem into more manageable subproblems.
3) Requires clear articulation of a research goal (i.e., the aim of the research).
4) Rests on certain critical theories.
5) Requires a specific plan to proceed.
6) Requires collection, analysis, and interpretation of data in an attempt to resolve the problem that initiated the research.
7) Is cyclical; there is no obvious end point because research encourages follow-up studies.

The term 'research' implies several elements: (i) the methodology, which includes the basic and critical theories, hypothesis, principles, and the logic of the research being undertaken; (ii) the research process and operational procedure; and (iii) detailed operational technique and tools for collecting and analysing data. The general research processes and steps are visualized in Figure 1.1. This diagram outlines a cyclical process of research, informed by literature and theories. With the problem being addressed through research, there is room for new research thinking to arise.

1.3 Cornerstones of Research

For research to be convincing and understandable to readers, it must follow certain research rules. In general, successful research has clearly defined and discussed elements.

1) Concept

A research concept is a set of meanings or characteristics associated with specific events, objects, conditions, situations, and behaviours in general, such as building energy consumption, organizational behaviour, corporate culture, and stakeholders. The success of research arguably depends first and foremost on the clarity of the concept and how

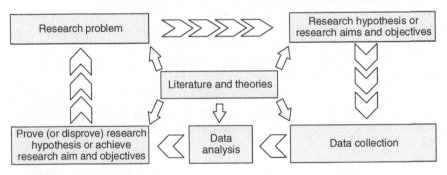

Figure 1.1   General research process.

others understand the concepts used. In some cases, the researcher may come up with a new concept, but this requires the researcher to define the concept clearly so that the reader understands the concept, and to maintain the strength and validity of the study.

2) **Theory**

   The Cambridge Dictionary defines theory is a formal statement of the rules on which a subject of study is based or of ideas that are suggested to explain a fact or event or, more generally, an opinion or explanation; this definition applies directly to research theory.

3) **Model**

   A research model is a formal representation of a practical problem or an object or law after abstraction. The difference between a model and a theory is that the role of a theory is to explain, whereas the role of a model is to demonstrate. A model is an important means of testing a theory.

4) **Construct**

   A research construct is an idea invented specifically for the purpose of a particular research or theory construction.

5) **Variable**

   In research, the variable is a specific value used to measure concepts or constructs.

6) **Proposition**

   A research proposition is a statement of concepts (variables) and relations between concepts (variables), used at the beginning of research design.

7) **Hypothesis**

   Hypothesis is a statement that attempts to explain phenomena or facts but has not yet been proven. Chapter 3, Section 3.1.1 expands upon this.

Apart from the above-mentioned cornerstones and elements, there are also other cornerstones of research, as shown in Figure 1.2. These elements and cornerstones are located in one of the three different phases of a research: conceptualization, operationalization, and measurement. Figure 1.2 also shows relevant components and methods which may be required to implement the research.

1.4 Philosophical Understanding of Research Methodology

It is important for researchers to clearly explain the philosophical assumptions that provide a foundation for the chosen research topic or problem before selecting a research methodology (Creswell and Clark 2017; Zou et al. 2014). Researchers' worldview (also known as paradigm) is the core of philosophical assumption. Thus, researchers need to be aware of the implicit worldviews they bring to their research (Creswell and Clark 2017). Worldviews directly affect assumptions the researcher makes about reality and the way to obtain knowledge (Creswell and Clark 2017).

There are four main types of worldviews applicable in research: positivism, postpositivism, constructivism, and pragmatism. Positivism believes that knowledge is based on natural phenomena, which is unbiased and cannot be affected by researchers' subjective view (Macionis and Gerber 1999). Positivism is associated with quantitative methods. As an amendment to positivism, postpositivism states that the subjective view of the researcher can influence

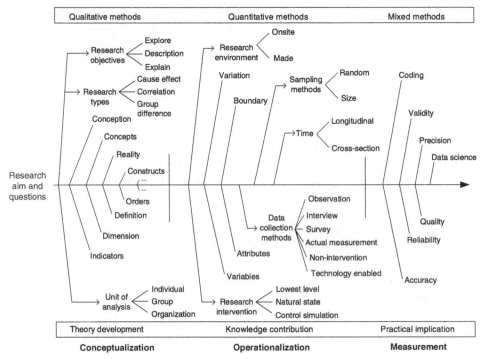

Figure 1.2   Relationships among cornerstones of research.

what is observed (Seaman 1995). To pursue objectivity, postpositivists use checks to recognize the possible effects of biases and eliminate them. Similar to positivism, postpositivism is also associated with quantitative methods. Contrary to positivism and postpositivism, constructivism is a worldview that is made up of the understanding and meaning of phenomena formed through the person being researched and their subjective view. It is believed that researchers and persons being researched are dependent on each other. Creswell and Clark (2017) pointed out that, in constructivism, researchers start with views of the persons being researched and build 'up' to patterns, theories, and generalizations. Constructivism is associated with qualitative methods. Due to the opposition of the worldview, there is a protracted debate between positivism (and postpositivism) and constructivism, which in turn has evolved into a conflict between qualitative and quantitative research methodologies. Under this background, pragmatism was born. Pragmatism is problem-oriented, believing that any method that can solve the research problem is a good method. With the support of pragmatism, mixed methods approaches have been developing rapidly in recent years.

To help readers gain an in-depth understanding of worldviews, five main philosophical considerations in research are discussed (Creswell and Clark 2017; Tashakkori and Teddlie 1998) below.

1) **Ontology** studies the nature of reality; researchers with different philosophical assumptions could have different views on the nature of reality. For example, positivism believes singular reality while constructivism believes multiple realities.

2) **Epistemology** explores the nature of knowledge, justification, and the relationship between cognition and reality.

3) **Axiology** examines whether researchers include biased perspective or not.
4) **Logical inference** focuses on the way to acquire knowledge, including induction and deduction.
5) **Rhetoric** concentrates on the language of research. For positivism, a formal style with clear definition of variables is used. In contrast, constructivism adopts an informal style, e.g. description.

Based on the above five philosophical considerations, researchers can recognize the philosophical assumptions that might underpin their particular research task, and use this to select a methodology, as proposed in Figure 1.3.

According to different worldviews, there are three common research methodologies: quantitative, qualitative and mixed methods, detail information is present in the following sections.

1.5 Qualitative Research Methods

Qualitative research emphasizes words and meanings rather than quantification in the collection and analysis of data (Bryman 2016). It develops interpretive narratives from the data in an effort to capture the complexity of those phenomena. Qualitative research tends to be subjective, where researchers begin with open minds and immerse themselves in the complexity of a situation before interacting with research participants. Data might be collected from a small number of participants, who are able to speak about the investigation topic. After sufficient data are collected, variables and theories are then drawn from the data, explaining the phenomenon in that particular context, which may or may not be generalizable (Leedy and Ormrod 2013). There are different qualitative research methods, such as ethnography, grounded theory, case study, phenomenology, narrative inquiry, and content analysis. Qualitative research is classified as a first research paradigm aiming at understanding a phenomenon. Readers who are interested in learning and applying these methods are encouraged to undertake further reading.

Chapter 2 discusses the definitions, steps, and methods for data collection and analysis in qualitative research. Particular emphasis is given to interviews as a method for data collection, including criteria for data sampling, because it is still a commonly used method. Emphasis is also given to data analysis methods, such as grounded theory, content analysis, and artificial intelligence (AI)-based qualitative analytical methods.

1.6 Quantitative Research Methods

Quantitative research typically tries to measure variables in a numerical way by using standardized instruments with the purpose to establish relationships among variables. This is the dominant methodology in natural science research and social science research. The process involves determination of concepts, variables, and hypotheses at the beginning of the research, which are tested after data have been collected. The data are collected from a population or from samples that represent the population so that research findings may be generalizable (Leedy and Ormrod 2013). The general process of quantitative research is shown in Figure 1.4.

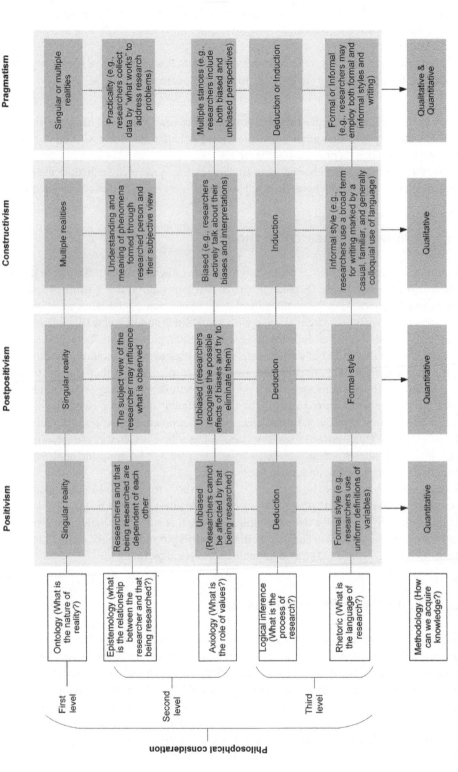

Figure 1.3 Methodology selection based on philosophical considerations.

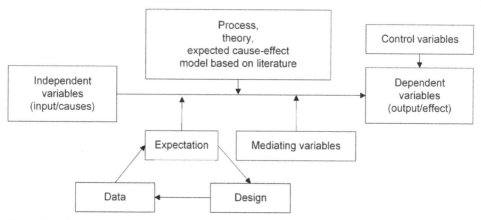

Figure 1.4 General process of quantitative research.

There are many methods of quantitative analysis, such as: statistical description, analysis of variance, meta-analysis, logic regression, multivariance analysis, correlation, factor analysis, principal component analysis, cluster analysis, nonparametric tests, and structural equation modelling. Chapter 3 discusses development of hypothesis, variables, and several methods of data analysis that are frequently used in quantitative research, including system dynamic approach, social network analysis method, interpretive structure modelling method, agent-based modelling, and data mining methods.

1.7 Mixed Methods Research Design

Mixed methods research design refers to a type of research that integrates quantitative and qualitative methodologies within a single research design. Many researchers believe that qualitative research and quantitative research methodologies complement rather than rival each other, and thus, qualitative research can compensate for the weaknesses of quantitative research and vice versa (Cooper and Schindler 2008). There are four aspects in deciding mixed methods research design: formative, paradigm debate, procedural development, and advocacy as separate design.

Bryman (2016) articulates that there are three approaches to mixed methods research: (i) triangulation: the use of quantitative research to corroborate qualitative research findings or vice versa; (ii) facilitation: one research methodology employed to aid research using the other research methodology; (iii) complementary: two research methodologies are employed so that different aspects of an investigation can be merged. However, there are also other methods of classifying the mixed methods. For instance, Creswell (2009) divided the mixed research into triangulation, embedded, explanatory, and exploratory through the relationship between qualitative and quantitative methods in the mixed research. The key to understanding mixed methods is to clearly articulate the relationships and functions of the two methods used in one single research design. Table 1.1 provides guidelines on the needs (when), the reasons (why), and the types (how) of using mixed methods research design.

Table 1.1 Guides for applying mixed methods research design.

Needs (when)	Reasons (why)	Types (how)
One form of data is insufficient by itself	To bring together the strengths of both QN and QL research to compare results or to validate, confirm, or corroborate QN results with QL findings	Triangulation design (convergence data transformation, validation of quantitative data, multilevel)
A second form of data is needed to enhance the study	There are different questions requiring different data	Embedded design (experimental correlational)
QN results are inadequate by themselves	QL data are needed to help explain or build on initial QN results.	Explanatory design (follow-up, participant selection)
QL results are inadequate by themselves	QL data are only an initial exploration to identify variables, constructs, taxonomies, or instruments for QN studies.	Exploratory design (instrument development, taxonomy development)

Note: QN – quantitative, QL – qualitative
Source: Adapted from Creswell, 2009.

Mixed methods research design is being used more and more as research problems and questions become increasingly complicated and complex, leading to the fact that one single research method is insufficient in solving the research problem or answering the research question. However, mixed method research design raises the challenge of finding the best way to mix the two different methods together and how to ensure the data collected are objective rather than subjective. There is also the potential that findings might be opposite or not complement one another. In addition, there may be concerns about the time-consuming nature of this method as it designs a mixture of qualitative and quantitative methods in many phases of the research process. Another important point for discussion is the indistinct boundary between social science research and natural science research under the mixed method research design. Mixed method research design arguably uses natural science methods to solve social science problems.

Chapter 4 examines mixed method research in great detail and depth, from definitions, mixed method research design process, and examples.

1.8 Technology-enabled Methods and Data-driven Methods

There are two main drawbacks of the research methods described above. Human bias may be embedded in the process of data collection and analysis. In addition, there is the time-consuming nature of these methods. With development of new emerging information and communication technology (ICT) and data science as well as artificial intelligence (AI), recent research has attempted to make use of these in research.

One main consideration is the use of ICT for data collection to cover the entire population or try to achieve objective data. For example, instead of asking how people act or behave in a given context, these actions or behaviour could be automatically captured and recorded using computer vision technologies (video cameras, scanners, etc.). Such automated data collection techniques and processes may allow the collection of a full set database over a certain period of time of the whole population. However, this process requires permission and cannot infringe upon privacy.

With the application of data science and AI, it is possible to analyse the big database to discover the cause–effect relationships between the variables through application of different algorithms. This new method overcomes the limitation of the paradigm of quantitative research methods, where hypotheses are normally put forward after literature review, and sample data are collected and analysed to approve or disapprove the hypothesis.

A technology- and data-driven methodological framework is shown Figure 1.5. Depending on the objective and scope of the research and the focuses and types of the research problems, research aims and questions are laid out. Thereafter, either qualitative, quantitative, or mixed research methods may be chosen. Based on the chosen method or methods, different technology-based data collection methods can be designed and implemented. After the data collection is completed, artificial intelligence, machine learning, or data science techniques can be used to analyse the collected data.

If following the process shown in Figure 1.5, the data collected and the results derived will be more objective than the traditional methods of data collection and analysis, due to the technology-based strategy of collecting data. Furthermore, technology- and data-driven research can help save time in data processing. The main implication of this framework is the need for enhancing researchers' technology knowledge and application skills, as well as data science-based data analysis skills.

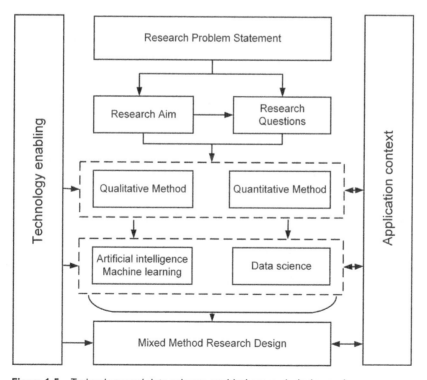

Figure 1.5 Technology and data science-enabled research design and process.

1.9 Theory Development

1.9.1 Definitions and Types of Theory

Before beginning to construct a theory, we need to have a clear understanding about the term 'theory' itself. Quite often, junior researchers or research students may get scared about the term 'theory' or 'knowledge contribution'. Theory is simply a statement about a phenomenon or a set of statements describing the interrelationships of the elements within a phenomenon – the 'what', 'why', or 'how'. Theory helps us understand phenomena, explaining what has happened in the past or predicting what might happen in the future. Theory is the starting point of deductive research and the end point of inductive research, and is an integral part of the scientific research process (Chen et al. 2008).

Generally speaking, theories are divided into three levels, namely grand, middle-range, and trivial theories (Yu et al. 2019). Grand theory is a highly complex, highly abstract, and systematic theory that attempts to encompass a large part of society, organizations, and individuals (Yu et al. 2019), such as scientific management theory and complex systems theory. Grand theories are more of a paradigm, representing common beliefs and perceptions in a broad sense.

Middle-range theory values the verifiability and observability of theories and requires the ability to find theoretical grounds in practice. The object of the middle-range theory concerns only a limited range of things and does not require an explanation of all phenomena. Middle-range theory lies between the grand theory and the trivial theory, and is the bridge between the two.

Trivial theory refers to common sense built up by ordinary people in their everyday lives. Unlike grand theory, trivial theory focuses on a limited number of concepts that are relevant to only a few phenomenon in a limited number of contexts.

1.9.2 Processes and Methods of Theory Development

Development of theory may arise from a process of conducting research, either qualitatively or quantitatively, to provide patterns, ideas, and ways of thinking for finding out what is happening with the phenomena or social events, or human relationships, or an economic phenomenon, or natural phenomenon. Steps of theory development via quantitative research might include:

1) Understanding the current theories for the given specific area of research.
2) Understanding the limits of the current theories.
3) Describing a phenomenon or a problem that cannot be explained by current theories, or is beyond the scope of current theories.
4) Formulating hypotheses or propositions that might describe or explain the phenomenon.
5) Testing the hypotheses or propositions by collecting and analysing data to prove or disapprove the hypothesis.
6) The tested hypotheses or propositions become a new theory explaining the phenomena from a particular perspective.

Alternatively, researchers could conduct qualitative research through observations, interviews, and review of literature to unfold the relationships and reasons behind a phenomena. When conducting qualitative research, a theory is developed towards the end of the research, as the result of the research.

Yet another method is case study research, where theory might be proposed before the case analysis and tested through case study, or developed after a case study implementation. The choice depends on which method is used, either induction or detection methods and perspectives.

In fact, it is not so difficult to develop a theory, as long as there is a scientific research process and research design, and the research is carried out according to this process and design framework. After all, research is about building on current understanding – this means researchers need to understand the current theory on the specific topic and then go beyond that current theory to make new contribution.

A few examples are provided here to demonstrate how a theory may be developed. The researcher should ask the right and important question as a starting point, then undertake research by observation, survey, interview, experiment, and so on to identify the internal relationship of the phenomena and develop and test hypothesis for further research. It is important to go beyond common sense, beyond the phenomena surface, particularly if there do not seem to be direct relationships but rather there are intermediate steps or hidden relationships between phenomena or within the elements of a phenomenon.

Taking safety management as an example, the researchers could ask how managers' skills may help improve safety performance in construction projects. The common theory covers three aspects of managerial skills: conceptual skills, human skills, and technical skills. But when thinking carefully, we know that there are more skill sets than that. For example, political skills are becoming more important. Therefore, one hypothesis could be that the four skill sets work together or individually to drive the improvement for safety performance in construction projects. Having set this hypothesis, one could go on undertaking research to prove or disapprove this hypothesis, which may become a revised or new theory once approved. The contribution would be linking the skill and safety and bringing political science into safety management domains.

Here is another example. A theory might be developed by analysing the cases of building energy retrofit projects. Looking into what was the driving factors for the success of building energy retrofit project development. Researchers might find out that a facilitation team was the key driving force bringing all parties and stakeholders together, and a new theoretical framework could be developed based on this finding, a framework articulating the importance of facilitation.

As another example, understanding complex risk interactions in project management has been recognized as an important research area, but how do researchers model this risk interaction network? Researchers could use social network analysis methods. The model may include risks occurring in different project stages or from different stakeholders. This requires complex systems knowledge and modelling and calculations. Researchers could also use social network analysis to see the key nodes in the network actions and how to simplify the network to simplify the risk interactions. The results of such research are a set of theories providing theoretical thinking and practical implications for managers as well as researchers.

The project risks could also impact on multiple project objectives; therefore, how to model such situations is another research that will lead to new theory and practical implications. This means we need to apply social network or complex system theories to conduct the modelling and simulation. Such research could also be extended into project life cycle risk management, capturing the reality that risk may happen in different stages of the project lifecycle. Again, there is a need to model how a risk that happens in one project stage may affect other stages. Having done such modelling and analysis, conclusions can be drawn and a phenomenon explained in creative ways. The outcomes of the research will form new statements, i.e., new theories explaining the above-mentioned project phenomena.

1.9.3   Inductive and Deductive Methods for Theory Development

Induction and deduction are widely used methods of logical thinking in scientific research. Marxist epistemology holds that all scientific research must use logical methods of induction or deduction. Induction is the method of generalizing the principles of a general conclusion from individual implementation, the method of moving from the individual to the general, the method of finding out the universal characteristics from empirical facts, and a method of summarizing scientific theorems, principles, or theories based on the accumulation of empirical materials. In the nineteenth century, Mendeleev used the inductive method to study the properties of the 63 chemical elements and the relationships between atoms, which led to the induction of the periodic law of the chemical elements and revealed the causal links between them.

In contrast to the inductive method, the deductive method starts from general premises and derives specific statements or individual conclusions. For example, according to the periodic law of the elements, Mendeleev anticipated not only the existence of new elements such as gallium, germanium, and scandium, which had not yet been discovered, but also the properties of these elements, which were successively confirmed by science.

Both inductive and deductive methods are often used for developing and testing theories in social research and management research.

1.10   Research Ethics

Ethics is 'conforming to the standards of conduct of a given profession or group' (Webster's new world dictionary). It refers to an agreement among members of a group, but different groups might have agreed on different codes of conducts. This means that what is ethical or unethical is relative to a particular society. Conducting research needs to be aware of the general agreements shared by all researchers and research participants about what is proper and improper in the conduct of specific research inquiries. Being ethical in research usually includes the following principles (Babbie 2020):

- Participation in research must be voluntary. In other words, no one can force the others to participate in the research. Participations include those directly involved or indirectly involved in any form of research, such as questionnaire surveys or face-to-face or phone-based interviews and face-to-face or online focused group workshops, etc.
- No harm to the participants. Under no circumstances should the research participants be harmed, either physically or mentally. Human decisions should never injure the people being studied.

- Information confidentiality. There is a need to keep all research information and data in a secure and confidential format to protect participants, particularly when there is sensitive and private information provided or gathered.
- Research processes should provide safeguarding to the participants so that they are not in any way placed in a dangerous situation or environment. All participants need to provide informed consent.
- Anonymity and confidentiality. This means the people who read about the research will not be able to identify a given response with a given respondent. This usually also includes the researcher himself or herself not being able to identify a given response with a given respondent. However, with the consensus of the participants, they can be identifiable, particularly when there is follow-up research to be carried out. Confidentiality refers to all information kept confidentially and the confidentiality agreement signed between the participants and researchers. The confidentiality of a research project is guaranteed when researchers can identify a given person's responses but promise not to do so publicly (Babbie 2020). Anonymity is achieved in the research project when neither the researchers or the readers of the findings can identify a given response with a given respondent.
- Both the researchers and participants must maintain honesty as part of their ethical standards. Lying about research purpose or giving forged answers are both unethical. This needs to be carefully considered in the research design. The actions taken by the researcher are ethical and the answers or responses provided by participants are true and complete. It is useful for the researchers to brief the research participants prior to the beginning of the research (Babbie 2020).

Researchers should also ensure ethical conduct during the analysis and reporting stages. To ensure am ethical process has been designed and implemented in the research, most institutions have an ethical review board or research ethics committees and researchers need to apply for ethical approvals. There is also professional code of ethics that is a useful reference and all research has to follow.

To safeguard the research, researchers must submit ethical approval applications to an institutional review board and obtain approval prior to the beginning of the research.

1.11 Research Impact and Citation

The journey of research has evolved from focusing merely on research design and outcome publication to a more longitudinal perspective that includes the impact and citation of research outcomes on theoretical and practical domains. This is partly due to the need to improve practice and partly driven by the advancement of information and communication technologies. As such it is necessary to plan and design a research project from a life-cycle longitudinal perspective, i.e., from research conceptualization and implementation to generating impact and improving citations. This also means researchers need to consider 'researching–writing–publishing' as a whole journey. Chapter 12 discusses this journey in detail, while Chapter 13 integrates all elements into a comprehensive framework.

1.12 The Book's Contents

Given the above discussions, this book includes two interrelated parts: research methodologies and research strategies (Figure 1.6). In specific, this book introduces several important topics of research methodology, including qualitative research (Chapter 2), quantitative research (Chapter 3), mixed method research design (Chapter 4), and case study research (Chapter 5). It also discusses new research methods using emerging technologies (Chapter 6) and data science (Chapter 7). Based on the discussions of different research paradigms, we provide a new outlook and define the fifth research paradigm, where social science research and natural science research intertwine together, becoming a complete new fifth paradigm (Chapter 8).

Following the research methods mentioned above, we have also included several chapters that are focused on research strategies, which cover how to write a manuscript for publishing in peer-reviewed international journals, from concept formation, idea development, and writing, to submission and responding to reviewer comments (Chapter 9). We also discuss thesis writing (Chapter 10). Furthermore, we have a chapter that discusses how to close the

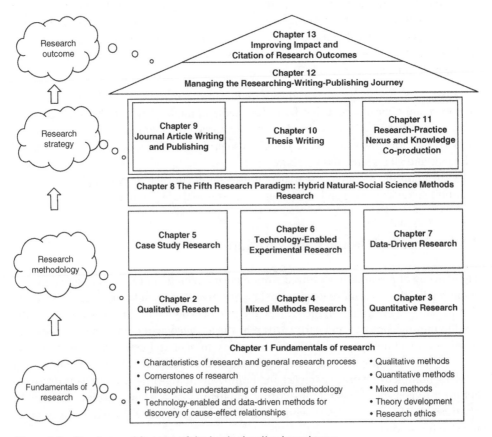

Figure 1.6 Structure and features of the book, visualized as a house.

gap between research and practice and offers a conceptual research-practice nexus framework (Chapter 11) based on the author's award-winning research outcomes. This chapter aims to enhance and improve research impact in practice.

Towards the end of the book, we provide a perspective on the researching–writing–publishing journey where we discuss the key points for formulating research ideas, strategies for producing high-quality journal articles, different thinking styles, and the relationship between thinking and writing (Chapter 12). We also provide an integrated framework that covers why conduct research, how to conduct research, and how to improve the impact of citation of research outcomes (Chapter 13), and a concluding remark that includes a clock-model of research critical success factors (RCSF) and the way forward into the future; this brings the book to a conclusion.

Numerous examples are provided for easy learning and understanding, application and practice of the concepts, methods, and techniques.

Some topics may appear several times across different chapters, in different perspectives and contexts. These repetitions also demonstrate the importance of these topics.

1.13   Unique Features of the Book

There are numerous unique features of this book, from methodological and strategic perspectives, with many examples and diagrams. Methodological features include: (i) discussion of advanced quantitative research methods, including system dynamics, agent-based modelling, social network analysis, interpretive structural modelling, and data mining with many application examples, drawing from our own experience and areas of expertise; (ii) discussion of mixed method research design, which includes the why, the what, and the step-by-step guidelines for readers to follow; (iii) discussion of the qualitative comparative analysis (QCA) method, which aims to integrate the advantages of qualitative and quantitative research methods in case study; (iv) development and application of technology-enabled experimental research methods and data-driven quantitative research methods, which are being applied increasingly in recent years, due to the improvement and maturity of information and communication technologies and availability and accessibility of massive data; and (v) development of the framework of the fifth research paradigm, where natural science research methods are integrated with social science research methods to form a completely new paradigm.

Strategic features include (i) discussion of theory development process and techniques, especially the theory classification, and inductive and deductive methods for theory development; (ii) strategies for writing, submitting, and publishing research outcomes, in specific, including paper structure and content, and techniques for responding to reviewers' comments, with numerous worked examples; (iii) comprehensive guidance on thesis writing, including overall design, innovation, structures, contribution to the knowledge, and originality, as well as the insight into thesis examination criteria and the pitfalls to be avoided; (iv) framework and discussion for enhancing research-practice nexus and knowledge coproduction, which improves the theoretical development and understanding and the ability to solve complex problems; and (v) life-cycle longitudinal framework for improving research impact and citations, including research conceptualization, implementation, and measurement as well as postpublication promotion and impact monitoring.

These are the contents that are currently lacking and we have filled in these gaps and provided useful information to researchers and practitioners in this book. Readers will enjoy learning and applying the new research methods, theories, and principles described in this book to improve their research productivity, quality, and outcomes – from research concept formation, process design, method reasoning, outcomes publishing, and impact generating.

1.14 Summary

This chapter has discussed the foundations of research by pin-pointing the characteristics of research, general research process, and cornerstones of research. It has also provided sufficient information to help readers understand research methodologies from a philosophical point of view, by linking ontology, epistemology, axiology, logical inference, and rhetoric. Different research methods have been briefly discussed to provide an overview that will facilitate the next several chapters, which are focused on research methodologies. A particular contribution of this chapter is the discussion of definitions and development of theory, which is the soul of any research. The chapter has also discussed research ethics, which is very important in today's research activities.

The figures presented in this chapter should be very helpful for readers to understand the above-mentioned topics and contents. Figure 1.2 shows the relationships among the cornerstones of research, conceptualizing the entire research process, research methods, theory, knowledge, and implication in more than 30 elements. Figure 1.3 brings together the four levels of philosophical consideration and different worldviews (paradigms) into the selection of research methods. Figure 1.5 shows how different research methods have evolved, been applied, and integrated. Figure 1.6 effectively presents the contents of this book and how the chapters are interrelated.

The 10 unique features of this book are also clearly explained. With an overview and discussions of the basic concepts and fundamentals of research, this chapter lays a solid foundation for other chapters presented in this book.

Review Questions and Exercises

1 Briefly describe the historical development of research methodology.

2 What are philosophical considerations in research?

3 What is worldview?

4 What role does data science play in research methodology?

5 What are technology-enabled methods?

6 What is the difference between quantitative and qualitative research methods?

7 What are the methods for developing a theory?

8 What are the requirements for research ethics?

9 What is the difference between induction and deduction methods?

References

Babbie, E.R. (2020). *The Practice of Social Research*, 15e. Cengage Learning.

Bryman, A. (2016). *Social Research Methods*. Oxford; New York: Oxford University Press.

Chen, X.P., Xu, S.Y., and Fan, J.L. (2008). *Empirical Methods in Organization and Management Research*. Peking University Press.

Cooper D. and Schindler, P. (2008). *Business Research Methods*, 10e. New York: McGraw-Hill/Irwin.

Creswell (2009). *Research Design: Qualitative, Quantitative, and Mixed Methods Approaches*, 3e. Thousand Oaks, CA: SAGE Publications.

Creswell, J.W. and Clark, V.L.P. (2017). *Designing and Conducting Mixed Methods Research*. SAGE Publishing.

Leedy, P.D. and Ormrod, J.E. (2013). The nature and tools of research. *Practical Research: Planning and Design* 1: 1–26.

Macionis, J.J. and Gerber, L.M. (1999). *Sociology – Third Canadian Edition*. Scarborough, ON: Prentice Hall Allyn and Bacon Canada.

Seaman, C.E. (1995). Real world research – a resource for social scientists and practitioner-researchers. *British Food Journal* 97 (10): 44.

Tashakkori, A. and Teddlie, C. (1998). *Mixed Methodology: Combining Qualitative and Quantitative Approaches*. SAGE Publishing.

Yu, X.Y., Zhao, H.D., and Fan, L.X. (2019). *Research Design and Methodology in Management*. China Machine Press.

Zou, P.X.W., Sunindijo, R.Y., and Dainty, A.R. (2014). A mixed methods research design for bridging the gap between research and practice in construction safety. *Safety Science* 70: 316–326.

References

2

Qualitative Research

2.1 Introduction

Qualitative research is concerned with words rather than numbers. In this chapter, we discuss the definitions and types of qualitative research and methods to conduct qualitative research, including methods and processes for theory development. We specifically discuss the interview, which is a popular data collection method in qualitative and mixed methods research. We also discuss artificial intelligent (AI)-based text mining methods.

There are three distinguishing features of qualitative research (Bryman 2016):

1) An inductive view of the relationship between theory and research, whereby the former is generated out of the latter. Inductive research may produce unanticipated findings based on the evidence gathered along with the explanation of its dynamics.
2) Research paradigm based on interpretivism, meaning that qualitative researchers believe that objects need to be investigated within the context of the real-world environment.
3) An ontological position described as constructionism, which implies that social properties are outcomes of the interactions between individuals, rather than phenomena 'out there' and separate from those involved in its construction.

Qualitative research originates from the humanities, specifically anthropology and sociology (Creswell and Creswell 2017). Its value and appropriateness in research has been increasingly recognized (Fellows and Liu 2016). As the attempts to address questions of how and why, qualitative research explores in-depth the manifestations of problems and issues that are the subject of quantitative studies and, thereby, to facilitate appreciation and understanding of basic causes and principles, notably, behaviours (Fellows and Liu 2016). The process of the qualitative research involves identifying emerging questions and procedures, data typically collected in the participant's setting, data analysis inductively building from particulars to general themes, and researchers interpreting the meaning of data, as shown in Figure 2.1.

2.2 Definitions of Qualitative Research

Literature defines qualitative research in one of two ways: in terms of what it is and in terms of what it is not. Qualitative research is a naturalistic and interpretivist approach to

Research Methodology and Strategy: Theory and Practice, First Edition. Patrick X.W. Zou and Xiaoxiao Xu.
© 2023 John Wiley & Sons Ltd. Published 2023 by John Wiley & Sons Ltd.

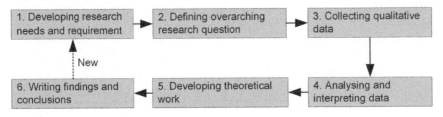

Figure 2.1 Main steps of qualitative research.

research that relies on participant observation, typically involving interaction between researcher and respondents. Thus, researchers attempt to gain an empathetic understanding of and in-depth insights into the experiences and opinions of the researched person (i.e., the respondents). It places emphasis and value on the human, interpretative aspects of knowledge about the social world and the significance of the researchers' own interpretations and understanding (Ritchie et al. 2013). In *The Handbook of Qualitative Research*, Denzin and Lincoln (1994) define qualitative research as: 'a situated activity that locates the observer in the world. It consists of a set of interpretive, material practices that makes the world visible. These practices turn the world into a series of representations including field notes, interviews, conversations, photographs, recordings, and memos. At this level, qualitative research involves an interpretive, naturalistic approach to the world. This means that qualitative researchers study things in their natural settings, attempting to make sense of, or to interpret, phenomena in terms of the meanings people bring to them.'

Qualitative research is concerned with words rather than numbers. Qualitative research may be considered the opposite of quantitative research, which is a positivist approach to research that relies on strictly scientific and highly structured methods, such as statistics, surveys, and questionnaires, attempting to yield representative and generalizable outcomes. Strauss and Corbin (1998) regard qualitative research as any research not primarily concerned with numbers or quantification of empirical observations: 'The term 'qualitative research' means any type of research that produces findings not arrived at by statistical procedures or other means of quantification.'

Ontologically, qualitative researchers agree that human beings are intricate and complex, and that human behaviour is not fully controlled by rigid external forces. Hence, it cannot be explained strictly by scientific laws. It is believed that the social world is governed by normative expectations and shared understandings and, hence, the laws that govern it are not immutable. Qualitative research is epistemologically associated with interpretivism, which claims that because the social world is not governed by regularities that hold law-like properties, a social researcher has to explore and understand the social world through the participants' and their own perspectives; and explanations can only be offered at the level of meaning rather than the cause (Ritchie et al. 2013).

Since qualitative research emphasizes words rather than quantification in the collection and analysis of data, three common characteristics of qualitative research are: using inductive reasoning to link theory and research; an interpretivist epistemological position that stresses the understanding of the social world through the interpretation of social participants; and a constructionist ontological position which views that social interactions influence social phenomena (Bryman 2016).

2.3 Types of Qualitative Research

There are several different types of research designs to conduct qualitative research.

2.3.1 Ethnography

Ethnography is the art and science of describing a group or culture (Fetterman 1998). In ethnography, it is common for the researchers to participate in the activities of the cultural group under investigation (participant observation). Ethnographic research methods are rooted in anthropological research and cross-cultural studies. In a broad sense, ethnographic research encompasses all studies of the social and cultural life of a particular group of people. The ethnographic research method is not a single method; rather, a variety of methods can be used to conduct ethnographic research. Ethnography aims to provide a holistic view and perspective on the understanding and interpretation of beliefs, attitudes, values, roles, and norms that arise in a particular socio-cultural context.

2.3.2 Grounded Theory

Grounded theory originated in the 1960s and was first proposed by two American sociologists, Barney Glaser and Anselm Strauss, in their monograph *The Discovery of Grounded Theory*. It is a systematic development of theory from data through inductive or deductive thinking (Phelps and Horman 2010). Grounded theory aims to derive a general, abstract theory of a social phenomenon grounded in the views of participants. Typically, it involves constant comparisons of data with emerging theories and theoretical sampling of different groups to maximize the similarities and differences of information. Grounded theory can be defined as 'a logically consistent data collection and analysis process for discovering theory, a research pathway that captures and conceptualizes underlying patterns in the social environment' (Glaser and Strauss 1967).

Grounded theory emphasizes the distillation of theory from information, arguing that theoretical frameworks are developed over time through in-depth analysis of information. Unlike the usual grand theories, rooted theory does not require the researchers to logically extrapolate predetermined hypotheses; instead, it starts with the information and analyses it inductively, resulting in a small theory that is applicable to a particular situation or problem.

It is important to emphasize that although it is common for researchers to think of grounded theory as a qualitative research method, it is possible to use quantitative data. Users of grounded theory can both quantitatively analyse qualitative information and qualitatively analyse quantitative data. Therefore, grounded theory is very advantageous in mixed methods research.

Grounded theory generally consists of eight steps: (i) identifying a problem or phenomenon to explore; (ii) collecting data; (iii) breaking transcripts into excerpts; (iv) grouping excerpts into codes; (v) grouping codes into categories; (vi) analysing more excerpts and compare with codes; (vii) repeating steps 2–6 until theoretical saturation is reached; and (viii) developing a theory or framework (Birks and Mills 2015).

2.3.3 Case Study

Case study is an idiographic examination of a single individual, family, organisation, event, activity, or process (Rubin and Babbie 2011). As a research method, case studies can be used in many fields. Case study allows researchers to improve their understanding of individuals, organisations, institutions, society, politics and other related fields. Case study has become a common method in sociology, psychology, political science, anthropology, social work, business and social planning. Researchers use the case study method to gain a comprehensive understanding of complex social phenomena. A variety of data collection methods can be employed to gain in-depth understanding concerning the case under investigation. Chapter 5 provides a more detailed examination of case studies.

2.3.4 Phenomenology

Phenomenology is a research design that aims to understand people's perceptions, perspectives, and understanding of a particular situation. A lengthy interview with people who have had direct experience with the phenomenon being studied is a typical method adopted in a phenomenology study (Leedy and Ormrod 2005). Phenomenology aims to elucidate specific phenomena through the perception of the phenomenon by actors in a given situation. In the human domain, this usually translates into gathering 'deep' information and perceptions through inductive, qualitative methods (such as interviews, discussions and participant observation) and understanding phenomena from the perspective of the research participants. Phenomenology is thus concerned with studying experience from the perspective of the individual. Specifically, phenomenological approaches are paradigms based on personal knowledge and subjectivity, and emphasize the importance of personal perspective and interpretation. As such, they are powerful for understanding subjective experience, gaining insight into people's motivations and actions, and penetrating the clutter of taken-for-granted assumptions and conventional wisdom.

2.3.5 Narrative Inquiry

Narrative inquiry is a study of the lives of individuals. The researcher typically asks one or more individuals to provide stories about their lives and then retells the stories in a narrative chronology that combines views from the participants and the researcher. This creates a dual layer of interpretation in narrative analysis.

Narratives may be drawn from a variety of sources, including diaries, letters, conversations, autobiographies, transcripts of in-depth interviews, focus groups or other types of narrative qualitative data. Narrative research generally consists of seven steps.

1) Determining initial research questions.
2) Selecting one or more participants to investigate.
3) Collecting stories from participant(s).
4) Retelling the participates' stories.
5) Working with the participates to ensure their experiences are accurately portrayed.
6) Writing the report and highlighting specific themes that emerged throughout the report.
7) Validating the report's accuracy (Daiute 2014).

2.4   Data Collection in Qualitative Research

Approaches to collecting qualitative data can be broadly divided into two groups: those that focus on naturally occurring data and those that generate data through the interventions of the research.

2.4.1   Naturally Occurring Data

There are numerous ways to identify and gather naturally occurring data. In participant observation, the researcher joins the constituent study population or its organizational or community setting to record actions, interactions, or events that occur. Observation also allows the recording of actions, interactions, and events but the researcher is not a member of the study population in this approach.

Documentary analysis refers to the study of documents, including media reports, government papers, or publicity materials, procedural documents like minutes of meetings, formal letters, or financial accounts; or personal documents like diaries, letters, or photographs. They are typically useful in gaining information about events or interactions that happened in the past, or when situations or events cannot be investigated by direct observation or questioning (Hammersley and Atkinson 2019). Conversation analysis is a detailed examination of 'talk in interaction' to determine how the conversation is constructed and enacted (Ritchie et al. 2013). The aim here is to investigate social intercourse, as it occurs in natural settings, in an attempt to describe people's methods for producing orderly social interaction (Silverman 2001). Discourse analysis is a combination of documentary analysis and conversation analysis that involves the study of constructed texts and verbal accounts that may include written documents, speeches, media reports, interviews, and conversation.

2.4.2   Generated Data

There are several approaches to generating data in qualitative research, including biographical methods, which use life stories, narratives, and recounted biographies; individual interviews, which are the most widely used method in qualitative research (Ritchie et al. 2013); paired or triad interviews, which allow more time for reflection, comparison, and interactive contemplation; and focus groups or group discussions, which involve between four and ten respondents that also allow for interaction, deeper reflection, and refinement of the respondent's initial insights and reactions.

Chapter 6 discusses different methods to use technology to generate and obtain objective research data.

2.5   Interview as a Method for Data Collection

The interview has long been used as one of the main research methods for data collection in qualitative research. This section provides a comprehensive overview of its definitions, types, sample selection, and sample size criteria. In particular, interview sample size determination criteria are discussed in great detail.

2.5.1 Definitions and Types of Interview

Interviews aim to elicit information about the interviewee's own behaviours, attitude, beliefs, norms, and values or that of others, by the interviewer from the interviewee. Rules of varying degrees of formality or explicitness help dictate the conduct of an interview (Bryman 2016).

An interview is a conservation with a structure and a purpose. It is not a normal daily conversation, in which equal partners share information, but a careful questioning and listening approach with the purpose of obtaining descriptions of the life world of the interviewees (Kvale and Brinkman 2009). There are different types and styles of interview design (Bryman 2016), including:

1) Structured or standardized interview: this entails the administration of an interview schedule, which is a collection of questions designed to be asked by an interviewer. The exact same questions will be asked of each respondent to ensure that the interviewee's responses can be aggregated.
2) Semi-structured interview: the interviewer has a series of questions in the general form of an interview schedule but is able to vary the sequence of questions. Also, the interviewer has opportunities to ask additional questions to probe further information, in response to what they see as significant replies.
3) Unstructured interview: the interviewer typically has only a list of topics or issues, often called an interview guide, that is to be covered. The style of questioning is usually informal. The phrasing and sequencing of questions will vary from interview to interview.
4) Focused individual interview: this interview uses predominantly open questions to ask interviewees about a specific situation or event that is relevant to them and of interest to the researcher. This can be administered to individuals or to groups. There are two analogous methods with the focus interview: *oral history interview* and *life history interview*. These are concerned with obtaining historical epoch or event information relating to each respondent. An oral history interview can be unstructured or semi-structured and asks interviewees to recall events from their past and reflect on them. Meanwhile, a life history interview, often unstructured, aims to obtain information on the entire of biography of each interviewee.
5) Focused group interview: this is the same as the focused interview, but interviewees discuss the specific issue in groups. There are several participants (in addition to the moderator/facilitator) with an emphasis in the questioning on a particular fairly tightly defined topic, and the accent is upon interaction within the group and the joint construction of meaning.

There are two considerations when justifying an interview design: research aim and constraints placed by interviewees. Regarding the first consideration, interviews vary in nature: structured, semi-structured, and unstructured. Table 2.1 lists the characteristics of interview types. Choosing an interview approach depends on the purpose of interview. If the researcher seeks specific information (e.g., for evaluation) without due regard for obtaining more detail, a structured interview can be used. In contrast, if the researcher requires in-depth information that reveals interesting and new aspects, an unstructured interview can be adopted. The semi-structured interview fills the spectrum between the two extremes. Furthermore, interviews can be carried out with individual or with a group of respondents, depending on whether the research seeks to elicit general consensus views of respondents or not (Fellows and Liu 2016). This leads to the second consideration, which is the

Table 2.1 Types and characteristics of interview.

Types	Characteristics
Structured	<ul><li>Data collected through formal style of questioning</li><li>Little scope for probing questions</li><li>Respondents choose an answer from alternatives</li><li>Same wording and questions for all interviewees</li></ul>
Semi-structured	<ul><li>Data collected through both formal and informal styles of questioning</li><li>Responses can be written and supplemented with recording</li><li>Responses are limited to subject in question, but interviewee is free to add more details if needed be</li><li>Provides more details than structured interviews about issues being investigated</li><li>Respondents provide topical answers</li><li>All respondents receive the same major issues</li></ul>
Unstructured	<ul><li>Data collected through informal style of questioning</li><li>Recording responses are most suitable</li><li>Respondents say as much as they wish after a brief introduction by the interviewer</li><li>They can be monologues with few prompts to ensure completion of statements</li><li>Respondents provide answers in any order they wish</li><li>Brief introduction of key issues to all respondents</li></ul>

Sources: Fellows and Liu (2016) *and* Bryman (2016).

constraint placed by interviewees; the researcher needs to consider the ability to gather a group of interviewees in one place for a focused group or group interview, or the interviewees willingness to participate in an expert panel for a Delphi method (a structured communication technique used to gain the opinions and insights of a group of experts or stakeholders on a particular topic or issue). The two concerns guide the selection of an interview design.

2.5.2 Sample Selection

Sampling in quantitative research and qualitative research is different (as shown in Figure 2.2). Qualitative researchers tend to emphasize purposive sampling, which is a nonprobability form of sampling (Bryman 2016). The researcher does not seek to sample research participants on a random basis, but to sample cases/participants in a strategic way, so that those sampled are relevant to the research questions that are posed (Bryman 2016). In deciding who or what to sample, reference should be made to what the research is about and whether those sampled will provide meaningful data to the research (Emmel 2013).

Figure 2.2 provides a flowchart of research sampling types. Purposive sampling can be divided into two approaches: theoretical sampling and generic purposive sampling (Bryman 2016; Emmel 2013). Theoretical sampling associates with grounded theory. Its definition is (Bryman 2016):

'[...] the process of data collection for generating theory whereby the analyst jointly collects, codes, and analyses the data and decides what data to collect next and where to find

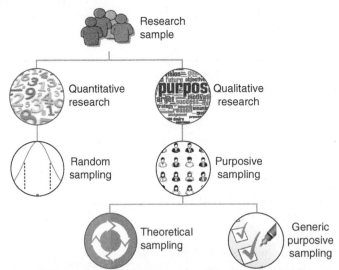

Figure 2.2 Types of research sampling. *Source:* Open Clip Art Library / Wikimedia Commons / CC0 1.0; skvoor/Adobe Stock Photos; laufer / Adobe Stock Photos; pking4th / Adobe Stock; HaywireMedia / Adobe Stock; mcmurryjulie / 109 images / Pixabay.

them, in order to develop the theory as it emerges. The process of data collection is controlled by the emerging theory, whether substantive or formal.'

The key criteria are that researchers should carry on sampling theoretically until a category has been saturated with data, meaning until (i) no new or relevant data seem to be emerging regarding a category; (ii) the category is well developed in terms of its properties and dimensions demonstrating variation; and (iii) the relationships among categories are well established and validated (Bryman 2016). Researchers decide when to cease collecting new data based on the emerging theory.

Different from theoretical sampling, generic purposive sampling is conducted purposively but not necessarily with regard to the generation of theory. The purpose of this sampling method is to select units that best provide insight into the research questions and will convince the audience of the research. Generic purposive sampling may be employed in a sequential or in a fixed manner and the criteria for selecting cases or individuals may be formed a priori (for example, sociodemographic criteria) or be contingent or a mixture of both (Bryman 2016). Purposive sampling refers to a set of nonprobability sampling techniques in which units are selected based on the characteristics that researchers need. It is designed before the research starts and may be redesigned as the research progresses, based on practical and pragmatic considerations driven forward by emerging theory (Emmel 2013). Sampling strategies for generic purposive sampling are shown in Table 2.2.

Decisions about whom or what to sample must relate to an understanding of the social world being investigated and the kinds of explanation the researchers are seeking to answer through the study (Emmel 2013).

2.5.3 Sample Size

While saturation is a core guiding principle to determine sample sizes in qualitative research (Hennink et al. 2017; Saunders et al. 2018), there are other considerations in deciding on the

Table 2.2 Sampling strategies for generic purposive sampling.

Sampling strategies	Description
Extreme or deviant case sampling	Sampling cases that are unusual or that are unusually at the far end(s) of a particular dimension of interest.
Typical case sampling	Sampling a case because it exemplifies a dimension of interest.
Critical case sampling	Sampling a crucial case that permits a logical inference about the phenomenon of interest – for example, a case might be chosen precisely because it is anticipated that it might allow a theory to be tested.
Maximum variation sampling	Sampling to ensure as wide a variation as possible in terms of the dimension of interest.
Criterion sampling	Sampling all units (cases or individuals) that meet a particular criterion.
Theoretical sampling	As discussed in previous sections.
Snowball sampling	Snowball sampling is a sampling technique in which the researcher samples initially a small group of people relevant to the research questions, and these sampled participants propose other participants who have had the experience or characteristics relevant to the research.
Opportunistic sampling	Capitalizing on opportunities to collect data from certain individuals, contact with whom is largely unforeseen but who may provide data relevant to the research question.
Stratified purposive sampling	Sampling of usually typical cases or individuals within subgroups of interest.

(Adapted from Bryman, 2016).

number of interviews. Malterud et al. (2016) proposed a concept called 'information power' to guide adequate sample size for qualitative studies. Information power indicates that the more relevant information the sample holds, the lower number of participants is needed. The size of a sample with sufficient information power depends on (i) the aim of the research, (ii) sample specificity, (iii) use of established theory, (iv) quality of dialogue, and (v) analysis strategy. Most of these factors are consistent with the five considerations when deciding how large a sample should be suggested by Bryman (2016), including (i) saturation, (ii) the minimum requirement for sample size, (iii) style and theoretical underpinning of the research, (iv) heterogeneity of the population, and (v) research scope and specificity. Similarly, Hennink et al. (2017) argued that many factors influence sample sizes for qualitative studies: the study purpose, research design, characteristics of the study population, analytic approach, and available resources. Synthesizing these factors produces a framework for justifying appropriate sample size for interview research, which is shown in Figure 2.3.

Six different aspects should be considered when deciding the sample size for interview research.

1) Saturation

Saturation is a concept of grounded theory proposed by Glaser and Strauss in 1967. In one form or another it now commands acceptance across a range of approaches to qualitative research (Saunders et al. 2018). Despite its popularity, a number of problematic conceptual

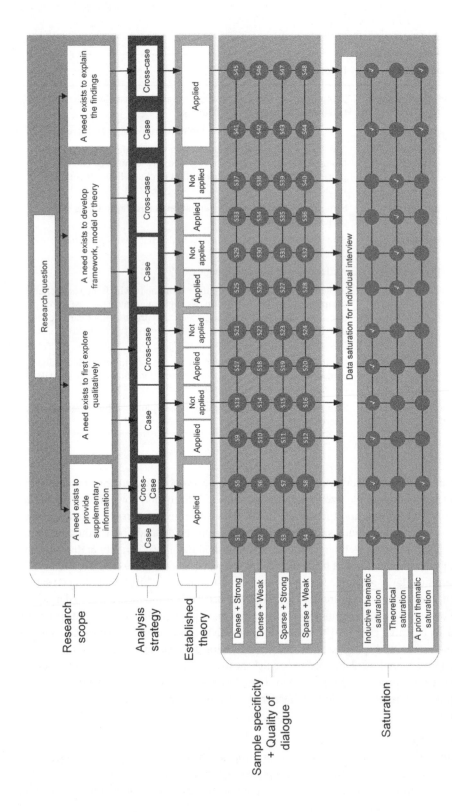

Figure 2.3 Framework for justifying sample size.

and methodological issues are raised in the literature (Saunders et al. 2018). For instance, little methodological research exists on parameters that influence saturation (Hennink et al. 2017). Saturation is often claimed (but not justified or explained) if criteria for deciding when theoretical saturation has been achieved are more or less absent (Bryman 2016). There are four modes of saturation (Saunders et al. 2018):

i) Theoretical saturation may be the most common used term by qualitative researchers. It is rooted in traditional grounded theory and uses the development of theory or categories in the analysis process as the criterion for ceasing data collection. Thus, it is concerned with the relevance of the sample in terms of testing emerging theoretical ideas (Bryman 2016).

ii) The second saturation mode focuses on the identification of new codes or themes and is based on the number of such codes or themes rather than the completeness of existing theoretical categories. This can be termed inductive thematic saturation. It can also be called 'information redundancy'. The research of Guest et al. (2006) and Hennink et al. (2017) reflects these types of saturation.

iii) The third mode is a priori thematic saturation, where data are collected so as to exemplify theory, at the level of lower-order codes or themes, rather than to develop or refine theory.

iv) The fourth mode is data saturation and aims to identify redundancy in the data within a specific interview, meaning that researchers need to continue asking probing questions until feeling that they have a full understanding of the participant's perspective.

Saturation should be considered a process, not a point. On this incremental reading of saturation, the analysis does not suddenly become 'rich' or 'insightful' after an additional interview, but presumably becomes *richer* or *more* insightful. The question is 'how much saturation is enough?', rather than 'has saturation occurred?' This is a less straightforward question but one that much better highlights that this can only be the analyst's decision – saturation is an ongoing, cumulative judgment, one that perhaps is never complete, rather than something that can be pinpointed at a specific juncture (Saunders et al. 2018).

2) Data analysis strategy

The strategy used for analysing research data relates to the sample size. Depending on the research aim and scope, researchers may attempt to conduct a cross-case analysis or an in-depth analysis of a few, selected participants. The cross-case analysis can be conducted in exploratory studies where researchers are keen on investigating a wide range of patterns associated with the research problem, while single case analysis is conducted to understand the different views of stakeholders. They may need to interview different groups of stakeholders and compare their perception together, resulting in a larger sample size. Other researchers may want to analyse why a stakeholder holds a certain view of a phenomenon. An in-depth analysis of data collected from participants can provide useful insights. Therefore, the analysis strategy is another factor affecting the sample size.

3) Quality of interview dialogue

A study with strong and clear communication between researcher and participants requires fewer participants to offer sufficient information, rather than a study with ambiguous or unfocused dialogues (Malterud et al. 2016). In a qualitative study, empirical data is coconstructed by complex interaction between researcher and participant, and a number of issues determine the quality of the communication from which the information power is established. Analytic value of the empirical data depends on the skills of the interviewer, the articulateness of the participant, and the chemistry between them.

4) Underpinning theory

A study supported by limited theoretical perspectives would usually require a larger sample to offer sufficient information power than a study that applies specific theories for planning and analysis (Malterud et al. 2016). A study starting from scratch with no theoretical background must establish its own foundation for grounding the results. Thus, larger sample size would be needed to grant sufficient information for meaningful analysis.

5) Heterogeneous population

In the case of a heterogeneous population, a larger sample may be required in order to reflect its inherent variability (Bryman 2016). The reason is that the sample size should be representative of the population (Boddy 2016). For a heterogeneous population, it is suggested that researchers draw up a grid (such as for different stakeholders in the building lifecycle) to make sure that each segment of the population is covered by the research (Wilson 1989). In contrast, there is a relationship between homogeneity of experience and smaller (Emmel 2013).

6) Research scope

A broad and general scope may require a larger sample in order to address the theoretical and empirical reach of the research questions (Bryman 2016; Malterud et al. 2016).

These six different aspects should be considered when deciding the sample size for interview research. Researchers should ensure that the sample size is not too small to make it difficult to obtain sufficient data to address the research question. At the same time, it should not be so large that it is difficult to undertake a deep, case-oriented analysis (Bryman 2016).

2.5.4 Framework for Selecting Types of Interview and Methods of Interview Data Analysis

Depending on the characteristics of the different research needs, we can choose different types of interviews (Figure 2.4). If a need exists to provide 'clear qualitative information', the researcher already has a clear direction for data collection and can, therefore, directly access the data they want to obtain through structured interviews. Semi-structured interviews may be used if the researcher also wants to obtain additional information that was not thought of beforehand, such as the reasons why certain information was generated and how it may have changed. In terms of selecting analysis methods, content analysis and text mining methods are suggested, as they can directly extract the quantitative and qualitative information the researcher wants.

Regarding 'a need to first explore qualitatively', researchers attempt to conduct an *exploratory analysis* of the research question and the research subject. In this case the researcher needs to gather as much qualitative data as possible related to the research question. Researchers could use unstructured interviews if there is sufficient time and researchers are ahead of interview control; otherwise, researchers should use semi-structured

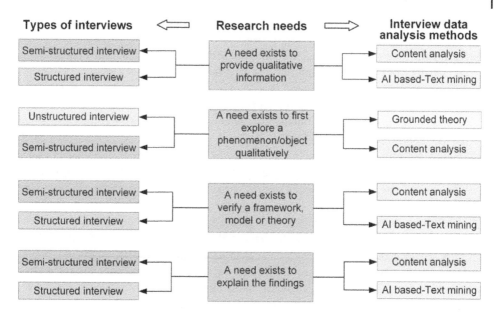

Figure 2.4 Framework for selecting types of interviews and methods for analysing interview data.

interviews. For data analysis, if researchers want to develop a theory based on the interview data, it is recommended to use grounded theory; if researchers want to explore or identify patterns, it is suggested to apply content analysis.

If a researcher needs to validate a framework, model, or theory, it is recommended to use structured or semi-structured interviews. The purpose of validating a framework, model or theory is straightforward and does not require much flexibility or freedom in the interview. Validation of a framework, model, or theory generally has more comprehensive and systematic criteria, where semi-structured interviews and structured interviews can be designed according to the criteria and compare expert perceptions and recommendations.

Regarding interview analysis, content analysis and new AI-based text mining methods could help systematically describe and uncover the meaning of qualitative data. Similarly, if a researcher needs to explain the findings, semi-structured or structured interviews can be used, where researchers design a series of questions around the findings of research and obtain respondents' responses to these questions. Content analysis and AI-based text mining methods can extract respondents' views and perceptions to explain the study findings.

2.5.5 Strategies for Conducting Interview

We put forward seven strategies for conducting interviews:

1) Researchers need to clearly define the research aim and objectives.
2) Write the questions to be used in the interview. These questions should cover the research aim and objectives, which will be achieved once the interview answers are analysed. It is always better to prepare more questions than not enough. Have at least

one open-ended question inviting interviewees to add anything that might have not been covered in the list of interview questions.

3) During the interview, follow-up questions are useful and may lead to surprise findings and results. For example, the interviewee may say something that researchers may not fully understand, or something new that researchers have not thought about. Having a follow-up question will allow researchers to get further information.

4) Analyse every interview once finished to understand the progress and the coverage of the research and if any modification on interview questions or process is needed. Having this done sooner will help improve the following interviews at the next or later stage. Continuous and prompt analysis of every interview will also help in deciding when to stop interviewing, by using saturation principles and other principles discussed in previous section.

5) Use the criteria presented in this book to decide when to stop interviews: when there is a sufficient number of interviews for data analysis and meaningful conclusions, and there are no abandoned interviews that consume too much time.

6) Use relevant software to analyse the interview data collected from the interview. It is necessary for the researchers to learn about how to use such software, e.g. Nvivo. Certain software can be helpful in transcribing recorded interviews. Researchers should use data science methods, such as natural language processing, content analysis, and machine learning.

7) The results and findings need to be compared with the current literature and current theory. Use common sense to justify if the research results and findings are reasonable, meaningful and useful. It is also important to think about whether the research results and findings make contributions to the current knowledge and theory, as well as practical implications and improvement.

2.6 Data Analysis in Qualitative Research

There are several approaches for data analysis in qualitative research. The following sections discuss analytic induction, grounded theory and content analysis as well as AI (artificial intelligent)-enabled text mining methods.

2.6.1 Analytic Induction

Analytic induction is the process by which a researcher defines a research question, proposes a research hypothesis, collects samples, examines them, and, if a deviant case is encountered, either redefines the hypothesis to exclude the problematic case or reformulates the hypothesis. The redefining or reformulating is conducted every time a deviant case is faced. Detailed information about case study can be seen in Chapter 5.

2.6.2 Grounded Theory

Grounded theory is one of the most widely applied methods of qualitative analysis. It refers to a set of procedures that include theoretical sampling, coding, theoretical saturation, and constant comparison. In this method, data collection, analysis, and eventual theory stand in close relationship to one another (Strauss and Corbin 1998). Two central features of grounded theory are that it is concerned with the development of theory out of data and that the approach is iterative, or recursive, meaning that data collection and analysis proceed in tandem, repeatedly referring back to each other (Bryman 2016). Makri and Neely (2021) provided a general guide for using ground theory for exploratory studies in management research. The overall processes and outcomes from grounded theory are:

1) Research question
2) Theoretical sampling
3) Collect data
4) Coding and continuous comparison of data
5) Saturate categories
6) Explore relationships between categories
7) Theoretical sampling
8) Test hypotheses
9) Collection and analysis of data in other settings.

2.6.3 Content Analysis Method

Content analysis is divided into quantitative content analysis and qualitative content analysis. Quantitative content analysis is a method for systematically classifying and recording the characteristics of textual, visual, or auditory material (Coe and Scacco 2017). It is unobtrusive in nature and does not rely on subjective perceptions (Krippendorff 2018). Qualitative content analysis aims to systematically describe the meaning of qualitative data (Schreier 2012). It helps to reduce the volume of material and requires the researcher to focus on selected aspects of meaning, i.e. those that are relevant to the overall research question (Schreier 2012). The use of content analysis method usually involves five steps:

1) Produce transcript of interview.
2) Reduce transcript of interview to coding categories that should be mutually exclusive.
3) Assign codes to the transcript of interview.
4) Check that the codes measure what the research intended to measure and check the consistency of the codes.
5) Identify patterns of and detailed information about codes.

2.6.4 Artificial Intelligence (AI)-based Text Mining Method

AI is increasingly being used to analyse research data. Within the many AI methods, text mining is most suitable for analysing qualitative data obtained through interviews. Text mining derives high-quality information from text data that are not easy to be revealed (Zhang et al. 2019). The commonly used text mining methods are word segmentation, sentiment analysis, and Latent Dirichlet Allocation (LDA).

Word segmentation is a method for dividing a string of written language into component words. The existing word segmentation algorithms can be divided into three main categories: lexicon-based methods, statistical-based methods, and rule-based methods.

Sentiment analysis is applied to systematically identify, extract, quantify, and explore affective states and subjective information (Prabowo and Thelwall 2009). Sentiment analysis mainly includes sentiment value calculation, sentiment orientation, and sentiment focus analysis. Sentiment value calculation is enabled by algorithms that assess the tone of segment words on a spectrum of positive to negative (Wang et al. 2019). Ekman's emotion model divides sentiment orientation into happiness, interest, sadness, anger, fear, surprise, and disgust (Tracy and Randles 2011). The first two indicate positive sentiments, while the rest indicate negative sentiments. Based on the word frequency with sentiment orientations, the sentiment focus of comments and users can be identified.

The Latent Dirichlet Allocation (LDA) topic model is the main method for topic modelling analysis. It is a document generation model and an unsupervised machine learning technique that considers a document as having multiple topics, each topic corresponding to a different word (Wu et al. 2021). LDA treats a document as a vector of word counts and a topic as a multinomial distribution of words (Wang et al. 2019). It finds the topics of this document and the words corresponding to topics by reversely simulating the document generation process.

2.7 Criticism about Qualitative Research

There are several criticisms about qualitative research. The use of limited samples to build an argument presents a weakness, particularly concerning the representativeness and generalizability of the research. Furthermore, critics argue that qualitative research lacks objectivity and tends to use personal opinions instead of evidence to support arguments (Grix 2004). Some critics also point out that a qualitative study is difficult to replicate because it is unstructured and often reliant upon the researcher's ingenuity (Bryman 2016).

2.8 Summary

In this chapter we have discussed the definitions and types of qualitative research, which included ethnography, grounded theory, case study, phenomenology, and narrative inquiry. We have also discussed the types of data and the methods for their collection. The interview as one of the major methods for data collection is discussed in detail, with emphasis on strategies for designing and undertaking interviews, in particular sample size determination criteria and strategies. We then discussed qualitative data analysis methods, including analytical induction, grounded theory, content analysis method, and artificial intelligence (AI)-enabled methods. We suggest that readers spend time to answer the review questions and do the exercises presented below, to articulate and help enhance your understanding of the key concepts, methods and techniques.

Review Questions and Exercises

1 What does the term qualitative research mean?

2 What are the methods can be used in qualitative research?

3 What types of methods are there in interviews as a cultivated data collection method?

4 What is the difference between grounded theory and content analysis?

5 What are the criteria to be used when determining the number of interviews in qualitative research, and how are these criteria applied?

6 What is the overall process of conducting QCA?

7 How may AI be used for data analysis in qualitative research?

References

Birks, M. and Mills, J. (2015). *Grounded Theory: A Practical Guide*, 2e. Los Angeles, CA: SAGE Publications.

Boddy, C.R. (2016). Sample size for qualitative research. *Qualitative Market Research: An International Journal* 19 (4): 426–432.

Bryman, A. (2016). *Social Research Methods*. Oxford University Press.

Coe, K. and Scacco, J.M. (2017). Quantitative content analysis. In: *The International Encyclopedia of Communication Research Methods*, 1–11. Hoboken, NJ: Wiley Blackwell.

Creswell, J.W. and Creswell, J.D. (2017). *Research Design: Qualitative, Quantitative, and Mixed Methods Approaches*. SAGE Publications.

Daiute, C. (2014). *Narrative Inquiry: A Dynamic Approach*. SAGE Publications.

Denzin, N.K. and Lincoln, Y.S. (1994). *Handbook of Qualitative Research*. SAGE Publications.

Emmel, N. (2013). *Sampling and Choosing Cases in Qualitative Research: A Realist Approach*. SAGE Publications.

Fellows, R. and Liu, A. (2016). Sensemaking in the cross-cultural contexts of projects. *International Journal of Project Management* 34 (2): 246–257.

Fetterman, D.M. (1998). *Ethnography: Step by Step*. Thousand Oaks: SAGE Publications.

Glaser, B. and Strauss, A. (1967). *The Discovery of Grounded Theory: Strategies for Qualitative Research*. Mill Valley, CA: Sociology Press.

Grix, J. (2004). *The Foundation of Research*. Basingstoke, UK: Palgrave Macmillan.

Guest, G., Bunce, A., and Johnson, L. (2006). How many interviews are enough? An experiment with data saturation and variability. *Field Methods* 18 (1): 59–82.

Hammersley, M. and Atkinson, P. (2019). *Ethnography: Practices and Principles*. New York: Routledge. Taylor & Francis Group.

Hennink, M.M., Kaiser, B.N., and Marconi, V.C. (2017). Code saturation versus meaning saturation: how many interviews are enough? *Qualitative Health Research* 27 (4): 591–608.

Krippendorff, K. (2018). *Content Analysis: An Introduction to Its Methodology*. SAGE Publications.

Kvale, S. and Brinkman, S. (2009). *Interview Quality. Interviews: Learning the Craft of Qualitative Research Interviewing*. Los Angeles, CA: SAGE Publications. 161–175.

Leedy, P.D. and Ormrod, J.E. (2005). *Practical Research: Planning and Design*. Upper Saddle River, NJ: Pearson.

Makri, C. and Neely, A. (2021). Grounded theory: a guide for exploratory studies in management research. *International Journal of Qualitative Methods* 20: 16094069211013654.

Malterud, K., Siersma, V.D., and Guassora, A.D. (2016). Sample size in qualitative interview studies: guided by information power. *Qualitative Health Research* 26 (13): 1753–1760.

Phelps, A.F. and Horman, M.J. (2010). Ethnographic theory-building research in construction. *Journal of Construction Engineering and Management* 136 (1): 56–65.

Prabowo, R. and Thelwall, M. (2009). Sentiment analysis: a combined approach. *Journal of Informetrics* 3 (2): 143–157.

Ritchie, J., Lewis, J., Nicholls, C.M., and Ormston, R. (2013). *Qualitative Research Practice: A Guide for Social Science Students and Researchers*. SAGE Publications.

Rubin, A. and Babbie, E. (2011). Research methods for social works. Brooks/Cole, Belmont, CA.

Saunders, B., Sim, J., Kingstone, T. et al. (2018). Saturation in qualitative research: exploring its conceptualization and operationalization. *Quality & Quantity* 52 (4): 1893–1907.

Schreier, M. (2012). *Qualitative Content Analysis in Practice*. SAGE Publications.

Silverman, D. (2001). *Interpreting Qualitative Data: Methods for Interpreting Talk, Text and Interaction*. London: SAGE Publications.

Strauss, A. and Corbin, J. (1998). *Basics of Qualitative Research: Grounded Theory Procedures and Techniques*, 2e. London: SAGE Publications.

Tracy, J.L. and Randles, D. (2011). Four models of basic emotions: a review of Ekman and Cordaro, Izard, Levenson, and Panksepp and Watt. *Emotion Review* 3 (4): 397–405.

Wang, Y., Li, H., and Wu, Z. (2019). Attitude of the Chinese public toward off-site construction: a text mining study. *Journal of Cleaner Production* 238: 117926.

Wilson, A. (1989). Qualitative market research: a practitioner's and Buyer's guide by W. Gordon and R. Langmaid. *Journal of Marketing Management* 5: 238–239.

Wu, Z., Zhang, Y., Chen, Q., and Wang, H. (2021). Attitude of Chinese public towards municipal solid waste sorting policy: a text mining study. *Science of the Total Environment* 756: 142674.

Zhang, F., Fleyeh, H., Wang, X., and Lu, M. (2019). Construction site accident analysis using text mining and natural language processing techniques. *Automation in Construction* 99: 238–248.

3

Quantitative Research

3.1 Introduction

Quantitative research is a very common research method consisting of two main steps, comprehensively collecting and mathematically analysing quantitative data. Like other types of research, quantitative research is centred around research problems, research aims, and research questions. One of the most important things in quantitative research is development of research problems, i.e., to propose research questions, and then develop and test the research hypotheses. Hypothesis development and testing is a method frequently used in quantitative research to evaluate the relationship between variables in theoretical models. There are many methods for analysing quantitative data, from traditional statistics to data mining. With the increasing complexity of research problems, several complex and systematic analysis methods are gaining attention from researchers. This chapter discusses hypothesis development and testing, and several commonly used analytical methods, including system dynamics, agent-based modelling, social network analysis, interpretive structural modelling, and data mining. Data-driven research methods will be discussed in more detail in Chapter 7.

3.2 Hypothesis Development and Testing

3.2.1 Development of Hypothesis

With a clear research question, the researcher can make the research problem specific through hypotheses (predictions of the outcomes of the variables). If existing research predicts a relationship between certain variables (for example, users with higher educational levels are more likely to engage in energy-saving behaviours), researchers will often use research hypothesis. In general, hypotheses are stated in one of the following formats: conditional statements, differential statements, statistical statements, and figurative statements.

1) Conditional statements can indicate whether two or more groups differ or are correlated under various variables. A conditional statement is expressed as follows: if A, then B.

Research Methodology and Strategy: Theory and Practice, First Edition. Patrick X.W. Zou and Xiaoxiao Xu.
© 2023 John Wiley & Sons Ltd. Published 2023 by John Wiley & Sons Ltd.

For example, if regulation of energy-consuming behaviour is enforced, users are more likely to engage in energy-saving behaviour.

2) Differential statements can be expressed in the form of 'A is different and B is different' or 'A is different and B is the same'. Differential statements consist of directional and nondirectional assumptions. The directional hypothesis compares whether there is a positive, negative, or size relationship between two variables, e.g., 'users with higher education are more energy efficient than those with lower education'. The nondirectional hypothesis describes the existence of a relationship between the two variables but does not specify the type of relationship, e.g., 'there is a difference in the perception of risk between men and women'. In general, the directed hypothesis is preferred because it is a clearer statement of the expected outcome.

3) Statistical statements are mainly used for statistical tests. Researchers can use statistical methods to verify the existence of hypothesized relationships. Statistical statements contain the null hypothesis and alternative hypothesis. The content of the null hypothesis is generally the assumption that one wishes to prove wrong. For example, in a correlation test the null hypothesis is usually 'there is no association between the two', while in an independence test the null hypothesis is usually 'there is an association between the two'. The null hypothesis is usually decided by the researcher and reflects the researcher's view of the unknown parameters, while the alternative hypothesis, which stands in opposition to the null hypothesis, needs to be proved to be true through statistical tests. The alternative hypothesis is accepted when there is sufficient evidence to reject the null hypothesis, while the null hypothesis is 'not rejected' if there is no sufficient evidence to prove that the alternative hypothesis is true.

4) Figurative statements transform words into causal patterns so that the reader can imagine the relationship between the independent, mediating, and dependent variables. For example, 'intention mediates between energy-saving behaviour and subjective norms'.

There are five different types of variables and their relationships are often present in hypothesis development and testing, namely independent variables, dependent variables, control variables, mediating variables, and moderating variables (Figure 3.1).

An *independent variable* is a factor or condition that is actively manipulated by the researcher and causes a change in the dependent variable. Thus, the independent variable

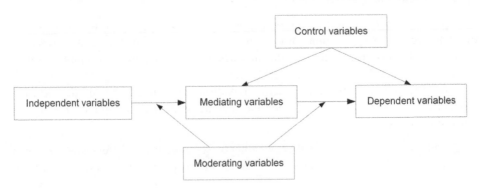

Figure 3.1 Types of variables and their relationships in hypothesis development and testing.

is seen as the cause of the dependent variables. A dependent variable is a variable that is a change in a phenomenon or an outcome because of a change in an independent variable; the independent variables and dependent variables are interdependent. For example, temperature and ice cream sales are a pair of independent and dependent variables where, as the temperature rises, ice cream sales will increase.

A *control variable* is a variable that has an effect on the dependent variable, and the effect must be excluded. Control variables are variables that researchers do not want in the study but cannot be randomized or eliminated. The *mediating variable* is the variable through which the independent variable influences the dependent variable and is the substantive, intrinsic cause of the influence of the independent variable on the dependent variable. *Moderating variables* are variables that affect the direction and strength of the relationship between the independent variables and dependent variables.

Each research subject is referred to as a unit of observation and a variable is a characteristic or attribute of the unit of observation, the specific value of the variable is referred to as the 'variable value'. Variables can be divided into qualitative variables or quantitative variables according to their type (Figure 3.2). Variables that take on qualitative values and are primarily code-bearers are qualitative variables, such as gender and educational level. Quantitative variables are quantitative, computationally meaningful variables, such as electricity consumption, age, and light intensity.

Quantitative variables can be divided into continuous and discrete variables. The range of values for continuous variables is theoretically continuous, for example, the range of values for building electricity consumption can theoretically be any positive real number. In contrast, the range of values of discrete variables is intermittent and discontinuous, for example, the number of building occupants, which can only be taken as an integer.

Qualitative variables can be divided into ordered categorical variables, unordered categorical variables, and grade variables. Variables that reflect a hierarchy or order, like educational level, are called ordered categorical variables. Ordered categorical variables differ in degree between the categories in which they take on values. Unordered variables take on values that do not differ in degree and can be divided into 'binomial categorical variables' and 'multinomial categorical variables' depending on the values taken. A binary categorical variable has only two values, e.g. biological gender (in general); a multinomial variable

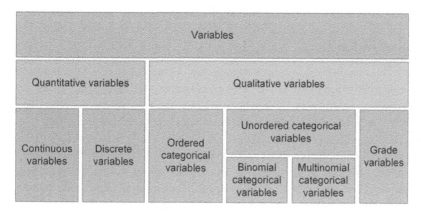

Figure 3.2 Classification of variables.

can have more than two values, e.g. religion, ethnicity, blood group. Grade variables are variables that allow things to be ordered in terms of the number or magnitude of an attribute of a thing, but without equal units. Grade variables are often used in questionnaires.

3.2.2 Design of Survey Questionnaire

Once the hypotheses for the research aim and objectives are developed, the next step is data collection. One of the unavoidable steps in quantitative research is to design a good set of questions for surveying, either in person or online. Then: What is a good set of questions? What are the criteria? Here are some pointers.

First, the questions should correspond to the research aims and objectives or propositions.

Second, the questions should be expressed as neutral as possible to avoid any subjectivity or bias.

Third, the questions should facilitate the respondents to open their minds.

Fourth, instead of using first-person style, it might be useful to use third person or properly explain the context of the research and questions so that the respondents do not feel like they are being questioned for their own situations.

Fifth, when designing questions for surveying, it is necessary to also think about how the answers of the questions will be analysed, and how the results of the surveys will lead to the achievement of the research aims.

Sixth, there should be an open-ended question to allow the respondents to provide their own opinions and comments.

Seventh, the questions should not be too long, but to the point and optimum; allow respondents to answer relatively quickly in a short time.

Eighth, the questions should reflect the interrelationships between the independent variables and dependent variables as well as mediating variables and control variables.

Ninth, the design of the questions should be based on current theories and proper reasoning process.

Tenth, the above criteria should be applied throughout the entire process of the research, from designing the questions to distribution of the questions and analysis of the questions.

3.2.3 Population and Sampling

The population is the theoretically specified aggregation of study elements, which means the group that researchers are interested in generalizing about (Babbie 2020). However, due to the limited time, resources, and cost, it is often not possible for researchers to collect all of the data from a population. Sampling is a selection of samples from a population. The basic requirement of sampling is to ensure that the samples are fully representative of the population.

The four common sampling methods include random sampling, systematic sampling, whole-group sampling, and stratified sampling. The random sampling method is a sample survey conducted in full accordance with the principle of equal opportunity. In a random sample, each part of the total population is equally likely to be selected. Systematic sampling is the process of arranging all units in the overall population in a certain order, randomly selecting one unit within a defined range as the initial unit, and then determining the other sample units according to predefined rules. Whole-group sampling starts by dividing the total into *i* groups, then randomly selecting a number of groups from the total number of groups and surveying all individuals or units within those groups. Stratified sampling is to divide the population into different types or strata according to their attribute characteristics, and then randomly select samples within the types or strata.

3.2.4 Testing of Hypothesis

Hypothesis testing is a statistical inference method used to determine whether the difference between samples, or a sample and the total, is caused by sampling error or by intrinsic differences. The basic idea of hypothesis testing is the principle of 'small probability events', whilst the method of statistical inference is the inverse method with some probability. The idea of small probability means that a small probability event will not occur in a single test. The counterfactual idea is to test the hypothesis first and then use appropriate statistical methods to determine whether the hypothesis is valid using the small probability principle. That is, in order to test whether a hypothesis H_0 (null hypothesis) is correct, it is first assumed that the hypothesis H_0 is correct, and then a decision is made to accept or reject the hypothesis H_0 based on the sample. If the sample observations lead to a 'small probability event', hypothesis H_0 should be rejected, otherwise hypothesis H_0 should be accepted (Taeger and Kuhnt 2014).

Significance testing is one of the most common methods of hypothesis testing and one of the most basic forms of statistical inference. This is done by making a hypothesis about the characteristics of the population and then an inference about whether this hypothesis should be rejected or accepted through statistical inference in a sample study. Commonly used hypothesis testing methods are z-test, t-test, chi-square test, F-test, and structural equation modelling, as discussed in the following sections.

1) z-test
The z-test is a test for differences in means for large samples (i.e., sample size greater than 30). Most random variables in real-world problems follow or approximately follow a normal distribution. The z-test is suggested for samples larger than 30 because, according to the central limit theorem, as the sample size increases, the sample is considered to approximately obey a normal distribution.

2) t-test
Different from z-test large samples the t-test is mainly used for normal distributions with small sample sizes (e.g. n < 30) and an unknown overall standard deviation σ. The t-test uses t-distribution theory to infer the probability of a difference occurring and thus compare whether the difference between two means is significant.

3) Chi-square test
The chi-square test examines the degree of deviation between the actual observed value and the theoretical inferred value of a statistical sample. The degree of deviation between the actual observed value and the theoretical inferred value determines the size of the chi-squared value. The greater the chi-squared value, the greater the deviation, and vice

versa; if the two values are exactly equal, the chi-squared value is zero, indicating that the theoretical value is exactly the same.

4) F-test

The F-test is a catch-all term for any test that uses the F-distribution. It is usually used to analyse statistical models that use more than one parameter to determine whether all or some of the parameters in the model are suitable for estimating the population.

5) Structural equation modelling

Structural equation modelling (SEM) is a statistical methodology that takes a hypothesis-testing approach to the analysis of a structural theory bearing on some phenomenon (Byrne 2013). This method allows a simultaneous examination of relationships among independent and dependent variables or constructs within a theoretical model (Mohamed 2002). The basic steps of SEM include specifying a model based on theory, determining how to measure constructs, collecting data, and analysing data which include overall model fit statistics and parameter estimates (Arbuckle and Wothke 2004).

3.2.5 Application Example

To demonstrate the process of hypothesis development and testing, we have chosen an example from Zou and Sunindijo (2013) 'Skills for managing safety risk, implementing safety task, and developing positive safety climate in construction project' (https://doi.org/10.1016/j. autcon.2012.10.018). The aim of this research is to understand what skills a project management team should develop in order to manage construction safety risks, implement safety tasks, and develop safety climate. A review of current literature was conducted to develop a hypothetical skill model. The structural equation modelling method was used to test the hypotheses and develop the skill model.

1) Research model and hypotheses

Construction project management personnel need to provide safety leadership in their projects. In order to provide this leadership, they need to manage safety risks and perform safety management tasks. By performing safety management tasks, project personnel contribute to the development of safety climate. This research argues that construction project management personnel's 4D skills (comprising conceptual, human, political, and technical skill), as illustrated in Figure 3.3, are what project personnel need to perform safety management tasks effectively.

Figure 3.4 demonstrates the input–process–output relationships between the three aspects. The project personnel's skills are the input and they influence the performance of the project personnel's safety leadership, which is expressed in the process form of the implementation of safety management tasks. When project personnel perform safety management tasks effectively, they will influence the output, i.e. development of safety climate in their projects.

The hypotheses that can be derived from the research model are:

Hypothesis 1: the higher the level of project management personnel's skills, the more effective the implementation of safety management tasks.
Hypothesis 2: the more effective implementation of safety management tasks, the higher the level of development of safety climate.

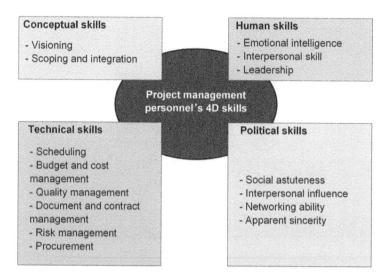

Figure 3.3 Construction project management personnel's 4D skills.

2) Structural equation modelling (SEM) analysis

The AMOS software (Version 18) was employed to undertake the SEM analysis for testing the hypotheses. During the testing process, it is found some paths are not significant with the standardized effect coefficients. After improving the hypothetical model by eliminating the insignificant paths, the model showed a good fit; Figure 3.5 shows the hypotheses testing results and the fitness indices are listed in Table 3.1.

To have a better understanding of the process of hypothesis development and testing, please read the following papers:

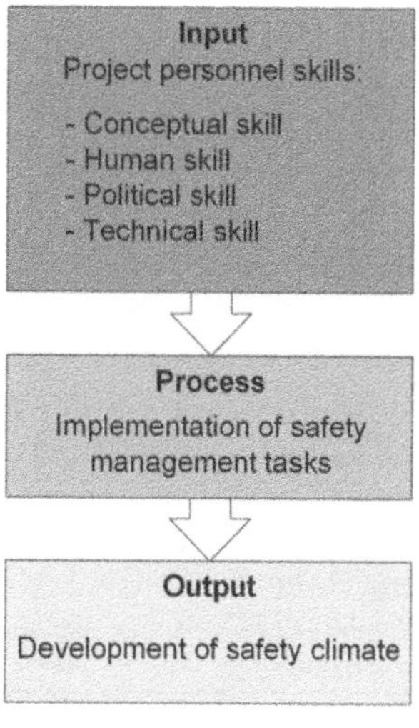

- Critical factors and paths influencing construction workers' safety risk tolerances https://doi.org/10.1016/j.aap.2015.11.027
- How safety leadership works among owners, contractors and subcontractors in construction projects https://doi.org/10.1016/j.ijproman.2016.02.013
- Why do individuals engage in collective actions against major construction projects? – An empirical analysis based on Chinese data https://doi.org/10.1016/j.ijproman.2018.02.004

Input

Project personnel skills:
- Conceptual skill
- Human skill
- Political skill
- Technical skill

Process

Implementation of safety management tasks

Output

Development of safety climate

Figure 3.4 Hypothesized research model (Zou and Sunindijo 2013 / with permission of Elsevier).

3.3 System Dynamics

System dynamics (SD), originated by J.W. Forrester, of MIT, in the 1950s, focuses on the structure of complex systems and the relationship between function and dynamic behaviour based on feedback control theory and computer simulation technology. It is useful not only for examining

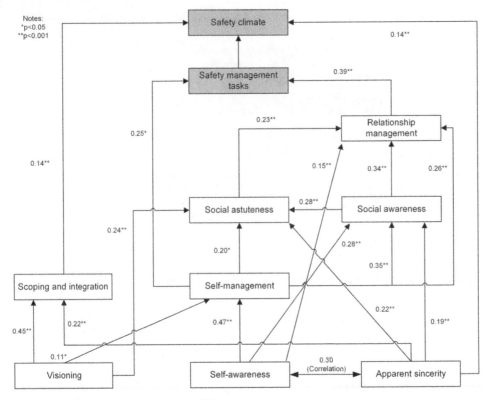

Figure 3.5 Skill components based on SEM analysis.

Table 3.1 Fit indexes.

Chi-square	RMSEA	GFI	AGFI	RMR	NFI	NNFI(TLI)	CFI
249.89	0.057	0.918	0.870	0.029	0.906	0.931	0.952
Degrees of freedom = 132							
p = 0.000							

the dynamic characteristic of a system but also for exploring the overall behaviour of a complex system that is difficult to predict (Yuan and Wang 2014). After nearly 60 years of development, SD has been widely used to deal with managerial, economic, environmental, and social systems of great complexity, such as economic development, military system management, energy and resources management, rural and urban planning, and construction management.

3.3.1 Model Boundary

Defining a clear model boundary is the first step in SD modelling. Factors or variables identification is the main method for researchers to define the model boundary. Factors or variables can be divided into three categories, i.e., endogenous, exogenous, and ignored (Ogunlana et al. 2003). Endogenous variables (or factors) are determined by the SD model, while exogenous variables (or factors) are determined by the factors outside the SD model. Ignored variables (or factors) can have an impact on the SD model but are not considered according to research aims. It is important to explicitly define endogenous variables, exogenous variables, and ignored variables. If the three kinds of variables are confused together, SD will not achieve the research aims (Sterman 2000). Literature review, workshop, interview, and specific process-based are three methods for researchers to define variables, and most of the time researchers combine some or all of these methods.

3.3.2 Model Development

There are two kinds of research using SD – qualitative analysis and quantitative analysis. Qualitative analysis uses only causal loop diagram. Causal loop diagrams (CLD) are drawn from a qualitative point of view. A CLD schematic is illustrated in Figure 3.6. As shown in the diagram, there are two kinds of arrow that represent a cause–effect relationship between two variables, namely positive (+) and negative (–). If the arrow-tail variable and the arrow-head variable change in the same direction, for example, the increase of A leads to an increase of B and the decrease of A leads to a decrease of B, then they have a positive correlation (Figure 3.6). Otherwise, they will have a negative correlation.

When the causal relationship constitutes a closed loop (the directions of arrows in the closed loop should be the same), there will be a feedback loop. The feedback loop is divided into two types: positive feedback and negative feedback. When the number of negative correlations in the feedback loop is an odd number, the feedback loop is a negative feedback loop. Otherwise, it will be a positive feedback loop (Figure 3.7).

When quantitative analysis is needed, the causal loop diagram needs to be converted into a stock-flow diagram (Yuan and Wang 2014). A stock-flow diagram contains four variables, i.e. stock, flow, auxiliary, and connector (Table 3.2).

An example stock-flow diagram can be seen in Figure 3.8. A stock variable, embracing tangible and intangible, shows the level of a system variable at a specific time (Ahmad et al. 2016), such as 'work to be finished' and 'completed work'. A flow variable, attached to a stock, measures the rate of change in stock, for example, 'processing rate' is the decrease in rate of 'work to be finished' and the increase in rate of 'completed work' (Figure 3.8). Therefore, the 'process rate' could directly influence the progress. As another flow variable, rework has the consequence of increasing the amount of work. 'Work to be finished', 'processing rate', 'complete work', and 'rework' form a feedback loop. An auxiliary variable serves as an intermediary for miscellaneous calculations (Li et al. 2014), e.g., 'fatigue', 'delay', and 'error'. They can be seen as the impact factor of 'processing rate' and 'rework'. A connector indicates connection and control between two variables (Ahmad et al. 2016).

3.3.3 Model Testing

According to testing standards, an SD model must pass the model structure test and model behaviour test. However, some studies focus only on the behaviour test and ignore the model structure test. It is worth noting that the model structure test and behaviour test are equally important. The consistency test of model behaviour and real system

Figure 3.6 Schematic of causality.

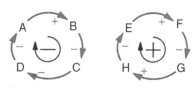

Figure 3.7 Schematic of feedback loop.

Table 3.2 Basic blocks used in system dynamics with icons.

Building block	Symbol	Description
Stock (level)		Shows a level of any variable in the system (Akhwanzada and Tahar 2012, Ahmad et al. 2016).
Flow (rate)		The rate of changes in stock and can cause the increase or decrease of a stock (Jin et al. 2016).
Auxiliary (convertor)		Connects stock and a flow in a complex setting. Used for intermediate calculations (Akhwanzada and Tahar 2012).
Connector		Denotes connection and control between system variables, showing the causality (Li et al. 2014).

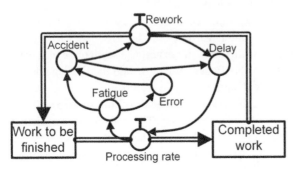

Figure 3.8 A sample stock-flow diagram (Xu and Zou 2021 / with permission of Springer Nature).

behaviour becomes meaningful only when the confidence of the model structure is established.

Structure tests include *Direct structure tests* and *Structure-oriented tests*. *Direct structure tests* assess the validity of model structure by comparing with knowledge about real system structure (Barlas 1996), while *Structure-oriented tests* assess the validity of model structure by applying behaviour tests on model-generated behaviour patterns (Senge and Forrester 1980). *Direct structure tests* include the structure-confirmation test, parameter-confirmation test, boundary adequacy test, and dimensional consistency test. The structure-confirmation test compares the causality and feedbacks of the model with the relationships that exist in the real system (Senge and Forrester 1980). The parameter-confirmation test indicates the evaluation of the constant parameters against the knowledge of the real system in terms of conceptual confirmation and numerical confirmation (Senge and Forrester 1980). Conceptual confirmation requires the parameters of the model to correspond to elements in the real system, and numerical confirmation requires sufficient accuracy of model parameters. The boundary adequacy test checks whether the model contains all important variables that affect the research objectives (Sterman 2000). The dimensional consistency test checks the right-hand side and left-hand side of each equation for dimensional consistency (Barlas 1996).

Structure-oriented tests mainly include three tests. The first is the extreme-condition test, which compares model-generated behaviour with the anticipated behaviour of the real system under some extreme conditions (Balci 1995; Barlas 1996). Another important structure-oriented test, integral error testing, checks whether the model behaviour will change significantly with the change in integration step (Sterman 2000). The third structure-oriented test, behaviour sensitivity testing means identifying the parameters to which the model is highly sensitive by observing the change in model behaviour through changing the variables in a reasonable range for identifying the sensitive variable (Barlas 1996). A sensitive variable must be highly accurate because its change will have an assignable effect on schedule.

A model behaviour test can be conducted to measure how accurately the model can reproduce the behaviour exhibited by the real system (Barlas 1996). Numerous methods are used to measure the accuracy of model behaviour, including R-square, Mean Absolute Difference, Mean Absolute Percentage Error, Mean Square Error, and Theil disequilibrium index. It is important to analyse the difference between model behaviour and real system behaviour rather than simply report how accurate the SD model is.

3.3.4 Model Simulation

Once researchers have built confidence in model structure and behaviour, the SD model can be used to conduct a simulation. Under normal circumstances, SD model simulation is mainly used for designing and evaluating improvement policies and strategies. As stated by Sterman (2000), policies and strategies design is not only changing model parameters but also creating new model structures. In addition, the interactions among different policies and strategies should be considered, as the impact of comprehensive policies and strategies is not equal to the simple sum of each policy and strategy (Sterman 2000).

3.3.5 Application Example

To demonstrate the application of the system dynamic modelling approach, we provide an example from Xu et al. (2018) 'Schedule risk analysis of infrastructure project: A hybrid dynamic approach', as presented in the following sections. For more details about the research and publication, readers are directed to read https://doi.org/10.1016/j.autcon.2018.07.026.

Schedule risk is a major concern in infrastructure project management. Existing studies have proposed several models for schedule risk analysis, but few efforts have been made on the dynamics and uncertainty of risks and the generality and practicability of the model. To fill the research gaps, this study develops a hybrid dynamic approach for investigating the effect of risks on infrastructure project schedule performance.

1) System boundary
The SD model in this study consists of five subsystems, namely: construction process, resource, project scope, schedule target, and project performance. As shown in Figure 3.9, each subsystem can be affected by schedule risks. Interactions among subsystems can occur, and the output, i.e., schedule delay, can be attained from schedule target subsystem.

2) Variables in the model
Considering the system boundary discussed in the previous section, the SD model uses three variables, namely, *input*, *output*, and *auxiliary* variables (Table 3.3). Variables were first collected from literature. Subsequently, they were cleared by collecting expert opinions in a series of interviews with 32 professionals. *Input* variables are all related to schedule risks and different schedule risks can be input into the SD model through *input* variable according to

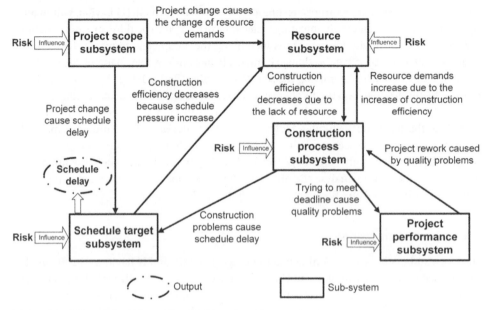

Figure 3.9 The relationship among the five subsystems.

the research objectives. The *output* variable is schedule delay. Auxiliary variables are those between input variables and output variables to facilitate the model simulation process.

3) Stock-flow diagram

There are 39 schedule risks and 15 corresponding input variables in the stock-flow diagram, as presented in Table 3.4.

A stock-flow diagram can be developed through the Vensim PLE® software package based on the previously discussed system boundaries and variables, as shown in Figure 3.10. Variables in red stand for the input variables, which can be affected by schedule risks. For example, design variations, as one of the schedule risks related to the designer may influence schedule through three variables: design change request, scope change rate, and delay due to approval procedures. 'Design variations' can increase the occurrence of a 'design change request', and the design change part may give 'scope change rate' (i.e. frequent changes in scope changes mean considerable work needs to be done). Meanwhile, 'design variations' may lead to 'delay due to approval procedures' because related stakeholders require sufficient time to approve design variations.

To have an in-depth understanding of the application of system dynamic modelling approach, please read the papers below:

- System-dynamic analysis on socio-economic impacts of land consolidation in China. https://doi.org/10.1016/j.habitatint.2016.05.007
- System dynamics analytical modelling approach for construction project management research: A critical review and future directions. https://link.springer.com/article/10.1007/s42524-019-0091-7
- Schedule risk modeling in prefabrication housing production. https://doi.org/10.1016/j.jclepro.2016.11.028

Table 3.3 Variables in the SD model.

No.	Abbrev.	Definition	Subsystem	Input-output-auxiliary type	Variable type in SD
1	ADC	Approved design change	Project scope	Auxiliary	Rate
2	ADCV	Actual design change variation	Project scope	Auxiliary	Rate
3	AP	Approval percentage	Project scope	Auxiliary	Converter
4	CSIR	Construction scope increase rate	Project scope	Auxiliary	Rate
5	DCPTC	Design change prior to commencement	Project scope	Input	Rate
6	DCR	Design change request	Project scope	Input	Rate
7	DCTBA	Design change to be approved	Project scope	Auxiliary	Stock
8	DDTAP	Delay due to approval procedures	Project scope	Input	Converter
9	RDC	Rejected design change	Project scope	Auxiliary	Rate
10	SCR	Scope change rate	Project scope	Input	Converter
11	TWQ	Total work quantity	Project scope	Auxiliary	Stock
12	DDTR	Delay due to reproduction	Project performance	Input	Converter
13	DDTRP	Delay due to rework process	Project performance	Input	Converter
14	DR	Defect rate	Project performance	Input	Rate
15	FR	Failure rate	Project performance	Input	Rate
16	HPTBDW	Hidden problems to be dealt with	Project performance	Auxiliary	Stock
17	QPTBDW	Quality problem to be dealt with	Project performance	Auxiliary	Stock
18	RR	Reprocessing rate	Project performance	Auxiliary	Rate
19	TR	Treatment rate	Project performance	Auxiliary	Rate
20	ANOW	Actual number of workers	Resource	Input	Converter
21	CE	Construction efficiency	Resource	Auxiliary	Converter
22	FD	Fatigue degree	Resource	Auxiliary	Converter
23	IWDTSD	Increased workers due to schedule delay	Resource	Auxiliary	Converter
24	ME	Mechanical efficiency	Resource	Input	Converter
25	MNOW	Maximum number of worker	Resource	Auxiliary	Constant
26	MQ	Material quality	Resource	Input	Converter

(Continued)

Table 3.3 (Continued)

No.	Abbrev.	Definition	Subsystem	Input-output-auxiliary type	Variable type in SD
27	NOM	Number of machines	Resource	Input	Converter
28	ONOW	Original number of worker	Resource	Auxiliary	Constant
29	PBEAP	Relationship between efficiency and proficiency	Resource	Auxiliary	Converter
30	R	Resource (includes material and machinery)	Resource	Auxiliary	Converter
31	RBFAE	Relationship between fatigue and efficiency	Resource	Auxiliary	Converter
32	RBPAE	Relationship between pressure and efficiency	Resource	Auxiliary	Converter
33	RM	Required material	Resource	Auxiliary	Constant
34	TEOM	Theoretical efficiency of machine	Resource	Auxiliary	Constant
35	WE	Work efficiency	Resource	Auxiliary	Converter
36	WH	Working hours	Resource	Auxiliary	Converter
37	WoP	Working pressure	Resource	Input	Converter
38	WP	Workers proficiency	Resource	Auxiliary	Converter
39	CW	Completed works	Construction process	Auxiliary	Stock
40	D	Duration	Construction process	Auxiliary	Constant
41	IR	Inspection rate	Construction process	Auxiliary	Rate
42	IS	Inspection suspended	Construction process	Auxiliary	Converter
43	PR	Processing rate	Construction process	Auxiliary	Rate
44	VW	Verified works	Construction process	Auxiliary	Stock
45	WS	Work suspended	Construction process	Input	Converter
46	WTBF	Works to be finished	Construction process	Auxiliary	Stock
47	CD	Current duration	Schedule target	Auxiliary	Converter
48	CP	Construction percentage	Schedule target	Auxiliary	Converter
49	RCTFID	Required current time for initial duration	Schedule target	Input	Converter
50	SD	Schedule delay	Schedule target	Output	Converter

Table 3.4 Schedule risks and their import interfaces.

No.	Type	Schedule risk	Input variable
1	Client	Clients' variation order	• Design change request • Scope change rate • Delay due to approval procedure
2	Client	Clients' slow decision making	• Delay due to approval procedure • Delay due to reproduction • Delay due to rework process
3	Client	Clients' cash flow problem	• Work suspended
4	Client	Clients' late contract award	• Work suspended
5	Client	Clients' poor management	• Delay due to approval procedure • Delay due to reproduction • Delay due to rework process • Work suspended
6	Client	Pressure due to tight project schedule	• Required current time for initial duration
7	Client	Over-high quality requirements	• Failure rate • Defect rate
8	Designer	Design variations	• Design change request • Scope change rate • Design change prior to commencement • Delay due to approval procedures
9	Designer	Incomplete drawing	• Work suspended
10	Designer	Inaccurate site investigation information	• Work suspended
11	Designer	Design drawings are hard to follow	• Work suspended
12	Designer	Deficiency in designs	• Work suspended • Failure rate • Defect rate • Delay due to rework process • Delay due to reproduction
13	Contractor	Contractors' financial difficulties	• Mechanical efficiency • Number of machines • Material quality • Work suspended • Actual number of worker
14	Contractor	Contractors' planning and scheduling problems	• Required current time for initial duration
15	Contractor	Contractors' inadequate site inspection	• Work suspended
16	Contractor	Shortage of manpower	• Actual number of workers

(Continued)

Table 3.4 (Continued)

No.	Type	Schedule risk	Input variable
17	Contractor	Unforeseen material damages	• Failure rate • Defect rate • Mechanical efficiency
18	Contractor	Equipment shortage	• Number of machines
19	Contractor	Equipment breakdown and maintenance problems	• Number of machines
20	Contractor	Lack of skilful labour	• Actual number of workers • Working pressure
21	Contractor	Failure in applying new technologies	• Work suspended
22	Contractor	Lack of safety insurance	• Work suspended
23	Contractor	Equipment delivery problem	• Number of machines
24	Contractor	Safety accidents	• Work suspended
25	Contractor	Material shortages	• Material quality
26	Contractor	Quality problem	• Delay due to rework process • Delay due to reproduction • Failure rate • Defect rate
27	Contractor	Slow mobilization	• Actual number of worker
28	Contractor	Interference with other trades	• Mechanical efficiency • Number of machines • Material quality • Work suspended • Actual number of worker
29	Subcontractor	Subcontractors' financial difficulties	• Mechanical efficiency • Number of machines • Material quality • Work suspended • Actual number of workers
30	Subcontractor	Delay in material provision due to subcontractors' fault	• Material quality
31	Government	Bureaucracy of the authority	• Work suspended
32	Government	Inappropriate interruptions by the authority	• Work suspended
33	Government	Complex official approval procedure	• Work suspended
34	Government	Unstable government policies	• Work suspended
35	External environment	Inflation	• Mechanical efficiency • Number of machines • Material quality • Work suspended • Actual number of workers

Table 3.4 (Continued)

No.	Type	Schedule risk	Input variable
36	External environment	Unsuitable weather conditions	Work suspended
37	External environment	Changes in rates of exchange,	• Mechanical efficiency • Number of machines • Material quality • Work suspended • Actual number of workers
38	External environment	Increase in oil price	• Work suspended
39	External environment	Contagious diseases	• Work suspended • Actual number of workers

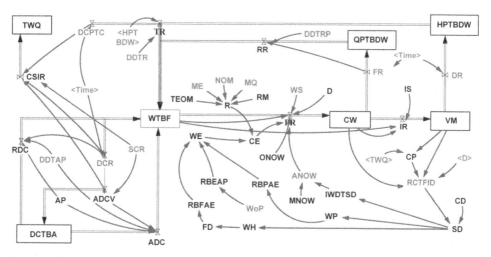

Figure 3.10 Stock-flow diagram.

3.4 Agent-based Modelling

Agent-based modelling belongs to the 'bottom-up approach', which concerns complex systems and simulates them through interaction between elements (Borshchev 2013; Iba 2013). An agent-based model is a kind of computational modes that simulates the behaviours and interactions of autonomous agents, e.g., stakeholder and organization, with a view to assessing their effects on a system.

3.4.1 Agent

An agent is independent, autonomous, goal-directed (Taylor 2014). Each agent can be distinguished from other agents by its attributes, while reacting to and interacting with other agents and the environment (Taylor 2014). An agent's behaviour can be affected by its attributes, other agents, or the environment. In agent-based modelling, agent behaviours are simulated based on *if-then-else* rules, complex artificial intelligence techniques or other models (e.g., system dynamics and discrete event simulation).

Every agent has cognition, which will help them examine their attributes and the environment around them to decide what actions to take. Russell and Norvig (1995) identified four types of agent cognition: reflexive agent, utility-based agents, goal-based agents, and adaptive agents. Reflexive agents react to input from other agents and the environment and take actions based on *if-then* rules. However, this kind of agent does not have the self-adjustment ability. Utility-based agents have utility functions and always attempt to maximize these utility functions. Goal-based agents have goals that they are using to dictate their actions (Russell and Norvig, 1995). For instance, the building occupants in public buildings will take actions to fulfil the goal of comfort. The agents with the above cognitions cannot change their strategy based on prior experience. To solve this problem, adaptive agents were created. Adaptive agents can learn from their past experience and change their behaviours based on their learning (Russell and Norvig, 1995). In this way, agents become more intelligent, in line with a real situation.

3.4.2 Agent Environment

The agent environment consists of the conditions and habitats surrounding the agents (Wilensky and Rand, 2015). The agent environment can affect agent decision and behaviour, and, in turn, can be affected by agent behaviours (Wilensky and Rand 2015). There are mainly two kinds of environment in agent-based modelling, namely spatial environments and network-based environments. In spatial environments, spatial information is an important attribute of agents. For example, spatial information of occupants in a commercial building is crucial for researchers to simulate occupant behaviour, as different spaces in a building have different microenvironments and each occupant may have different behaviour. In the network-based environment, agents' interactions are not defined by individual communications but rather by physical geography.

3.4.3 Interactions

There are four kinds of interactions in agent-based modelling, including agent–self interactions, agent–agent interactions, environment–self interactions, and agent–environment interactions (Wilensky and Rand 2015), as shown in Figure 3.11.

In an agent–self interaction, an agent interaction is done within itself. That is, an agent may decide to do something based on its state rather than the interactions from the environment or other agents. Sometimes, the environment may alter and change itself; for instance, the outdoor environment has impact on indoor environment. Agent–environment and agent–agent interactions are the main components of agent-based modelling. Agent–environment interactions happen when the agent take actions that have impact on the environment, or when the environment has changes that impact on agents. In agent–agent interactions, one agent takes actions based on the behaviours of other agents.

3.4.4 Application Example

The above-mentioned method has been used in the authors' previous research 'Agent-based model for simulating building energy management in student residences' (https://

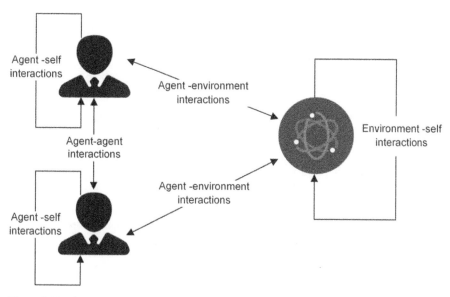

Figure 3.11   The four types of interactions in agent-based modelling.

doi.org/10.1016/j.enbuild.2019.05.053). Reducing energy consumption in buildings through behavioural changes has been regarded as a relatively low-cost and sustainable method. However, studies that focus on occupant building energy consumption in student residences are few. In the context of shared residences, the energy behaviour could be very different. This is because student–student and student–building system interactions are complex. To address this research gap, this example developed an agent-based simulation model regarding students as heterogeneous individuals.

1) Agent identification and assumptions
Agent identification: (1) student, (2) dormitory, (3) air-condition, (4) light, (5) computer, (6) facility manager, (7) education department, and (8) dormitory building agents.

Basic assumptions: (1) only dormitory energy consumption-related agents and variables were considered; (2) only dormitory electricity consumption was covered, and water consumption was disregarded; and (3) the model simulation time was from 1 January 2017 to 31 December 2017 with a time step of one hour.

2) System identification and decomposition
From the perspective of complexity science, the energy consumption management system is a complex adaptive system. Based on literature review and empirical data analysis, the energy consumption management system should include the energy consumption subsystem and the energy consumption supervision subsystem. The type of agents included in the subsystems and the relationships between the subsystems are illustrated in Figure 3.12.

The energy consumption subsystem included students and equipment. The students were the direct stakeholders who consumed energy. Their energy consumption pattern, energy-saving awareness, understanding of energy-saving knowledge, and daily energy-saving behaviour might affect energy consumption. Primary equipment included air conditioning, computer, and light. Therefore, these three types of equipment were considered in this research.

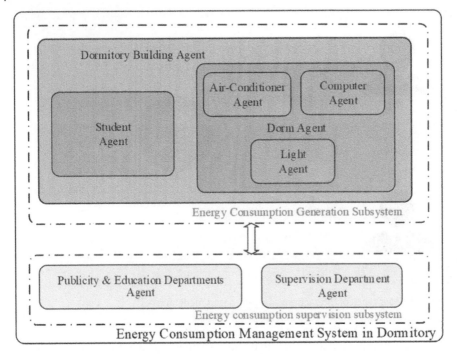

Figure 3.12 System boundary for energy consumption management.

The energy consumption supervision subsystem covered the logistics supervision department and the propaganda and education departments. The university logistics supervision department directly affected the energy consumption in the operating stage. This department often took measures, including energy consumption monitoring, promulgating the energy-saving regulations and energy-saving transformation. The concept or behaviour of students was affected by publicity and education departments in universities (e.g. student unions and community organizations) to a large extent.

The analysis object, attribute, behaviour, and knowledge of each agent in the energy consumption management system of dormitories and the design of the conceptual model of each agent are shown in Figure 3.13.

3) Model specification

After analysing the system and agents and their relationships, interactions, and behaviours, concepts must be formulated to be computer implementable. Model formalization was needed, which described in detail the behaviour rules of various subjects in the model according to the actual research. On the basis of object-oriented programming with Java, a system simulation model for dormitory energy management, including main function, student, air conditioning, computer, light, dormitory, dormitory building, education department, and supervisor, was developed in AnyLogic platform, as shown in Figure 3.14. Parameters, variables, functions, events, and statecharts were used to define the attributes and behaviour rules of agents.

Each type of agent had a statechart, including parameters and variable events. The student agent was presented as examples to illustrate the construction of agents. Relevant information for building the student agent was determined by analysing the students. The

Student Agent
- Object: With minimal financial burden to get a comfortable living environment.
- Attribute: Student basic information, the status of at school or leaving school, activities of student at school, habits of usin g appliances, statistic time of students in the room, awareness level of energy-saving.
- Behavior: Go to class, get up, sleep, turn on appliances, turn off appliances, take part in activities, get out of school and so on.
- Knowledge: Random movement within the Moore neighborhood in two dimensional space, affected by the influence of other agents in the neighborhood (limited rationality), such as affected by the surrounding students or affected by propoganda education to improve the level of energy cionservation awareness; reduce energy consumption with the improvement of energy saving consciousness.

Light Agent
- Object: Statistics of the use of light and calculation of energy consumption.
- Attribute: Power, switch, energy consumption and usage time of Light.
- Behavior: Determine whether the electric light is running, and count the daily usage time and energy consumption of the electric light.
- knowledge: The state of the light which passed by the student agent is to judge whether it is running. If it is running, record the opening time of the light and calculate the energy consumption.

Air conditioning Agent
- Object: Statistics of the use of air conditioning and calculation of energy consumption.
- Attribute: Setting temperature, power, energy consumption and usage time of Air conditioning.
- Behabior: Determine whether the electric light is running, and count the daily usage time and energy consumption of the air conditioning.
- Knowledge: The state of the sir conditioner which passed by the student agent is to judge whether it is running. If it is running, record the opening time in time and calculate the energy consumption according to the setting temperature of the air conditioner.

Computer Agent
- Object: Statistics of the use of computer and calculation of enrgy consumption.
- Attribute: operate power, standby power, running time, standyby time, energy consumption of operating and standby.
- Behavior: Determine the status of computer, count daily computer usage time, standby time and energy consumption.
- Knowledge: The state of the computer which passed by the student agent is to judge whether it is running. If it is running, record the running time in time and calculate the energy consumption.

Supervisor Agent
- Object: To ensure daily life and work of student, to pursue the greatest social and environmental benefit.
- Attribute: Daily energy consumption of female dormitories, male dormitories and all the dormitories.
- Behavior: Implement night electricity control and count the total energy consumption of all dormitories.
- Knowledge: Determine whether today is the school day at 0:00 every day, transmit the message of lights out to all the dormitory to achieve the effect of electricity control.

Dormitory Building Agent
- Object: Import basic information, statistics of the number of students in various categories.
- Attribute: Dormitory buildings of male and female.
- Behavior: Dormitory assignment, import basic information, vacation information, and course information of student.
- Knowledge: Information required to import the model based on actual school data.

Education Department Agent
- Object: Adopt various methods of publicity and education
- Attribute: Publicity coverage rate, audience number, publicity activities, participation rate, participation number, appraising activities.
- Behavior: Organize various promotional activities
- Knowledge: With the innovation of activities, the number and rate of participation increased, increasing awareness of energy-saving

Dormitory Agent
- Object: Statistics of the number of dormitory and calculation of energy consumption.
- Attribute: Dormitory number,student1,student2,student3,student4,the total daily energy consumption of the dormitory, light, computer and air conditioning, the number of students in the dormitory.
- Behavior: Statistics of the number of students in the dormitory, electrical running time and energy consumption
- Knowledge: Real-time statistics on the running time, operating energy consumption and number of students in the dormitory. The dormitory plays a role of bridge between the students and the appliances.

Figure 3.13 Conceptual model for the management of dormitory energy consumption (Ding et al. 2019 / with permission of Elsevier).

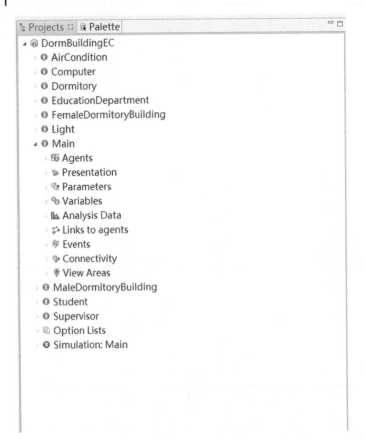

Figure 3.14 Simulation model for managing dormitory energy consumption.

student agent included the basic information of students, the status of whether students were in dormitories, the behaviour when using appliances, and the awareness level of energy saving. Differences in student awareness of energy conservation affected behaviour when using electrical equipment. Student behaviour also crucially affected energy consumption. The occupancy of dormitories was influenced by curriculum information. Equipment status was influenced by many factors, and the influence factors of different equipment were different. The student agent model diagram was built in the AnyLogic platform on the basis of the preceding analysis and parameter definitions. Three modules, i.e. parameter setting, statistical function, and behaviour control, existed.

Parameter, function, and event are shown in Figure 3.15. Various parameters were set in the model to reflect student occupancy, course information, and equipment information to simulate the activities of students realistically. The event was realized by function. Parameter settings included (a) student information, which contained the basic information and awareness of students, (b) status of staying at dormitories, which was influenced by course, the probability of staying in dormitories during holidays, and (c) behaviour when using appliances.

The parameter of 'SavingConsciousness' determined whether electrical appliances are turned off in time. Based on the questionnaires, students were affected by publicity and the behaviour of other students. Hence, the parameter of 'AcceptanceRate' was also set as the willingness to accept information on energy saving. The influence of other students' behaviour could be positive or negative. The parameter of 'NegativeInfluence' means the influence which resulted in more waste than before. Various parameters were set in the model (Figure 3.15) to reflect student occupancy, course, and appliance. Some parameters are shown in Table 3.5.

Behaviour control included movement, equipment, and awareness control (Figure 3.16). Figure 3.16(a) shows the primary activities of a student in a university, such as sleeping, waking up, going to class and going out, with the red dotted line marking the part that controls the students' sleeping and waking. The blue dotted callout section indicates that students were required to attend classes on the same day, ranging from top to bottom to

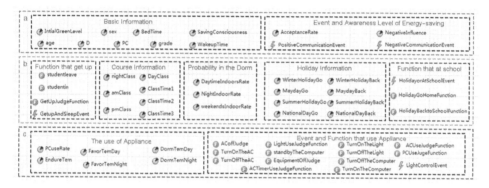

Figure 3.15 Parameter settings for students in the model (Ding et al. 2019 / with permission of Elsevier).

Table 3.5 Parameters of the student agent.

Parameter name	Parameter type	Meaning
PC	Computer	Personal computer
Wake-up time	double	Time of waking up in the morning
SavingConsciousness	double	Awareness of energy saving and environmental protection
InDormTime	double	Time in the dormitory
SleepTime	double	Time of sleeping
Dayclass	int	Whether a student had a class on that day
ClassTime1	double	Number of courses in the morning
WeekendsIndoorRate	double	Probability of staying in the dormitory on weekends and holidays
PCuseRate	double	Proportion of time students spend using computers in dormitories
Favourite day	int	Personal air-conditioning comfort temperature during the day
Domesday	int	Dormitory daytime air-conditioning setting temperature

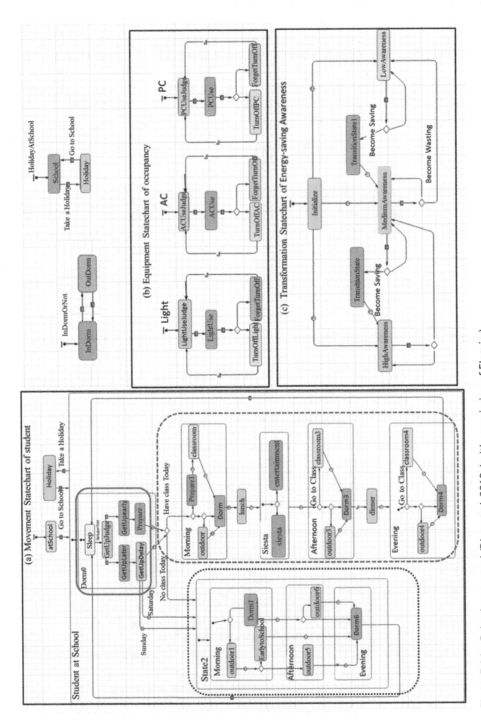

Figure 3.16 Behaviour control (Ding et al. 2019 / with permission of Elsevier).

four sections in the morning, noon, afternoon, and evening. The black dashed line indicates the status chart of students who had no class hours during the weekend or the entire day, which was divided into morning, afternoon, and evening. Figure 3.16(b) shows the statechart that reflected the occupancy of appliances, including light, air conditioning, and computer. The principle of 'judging before implementation' was followed whilst turning on or off appliances in the model. When students entered a certain state or were in the transition path amongst different states, code or function would be executed to pass specific messages to the target device to control the appliances.

Figure 3.16(c) depicts the state that reflected the transformation of students' awareness level of energy saving. The preliminary clustering analysis showed that students were divided into three groups, namely, high, medium, and low energy conservation consciousness levels. During model initialization, in accordance with the proportion of the three types of students in the real world, energy conservation consciousness would be initialized. Students would receive information on energy saving from others, in which students would have a certain probability of changing their energy-saving awareness, e.g. from 'Low Awareness' to 'Medium Awareness.' The probability was controlled by the parameter 'Acceptance Rate', which was from the questionnaire. Real information, such as basic, course, and weather information and parameters, was imported into the model. The dormitory energy consumption in this simulation model was collected from a University. Other empirical data were collected through questionnaire investigation and literature review, thus combining quantitative methods (agent-based modelling) with information gained through qualitative research.

To have a better understanding of ABM, please read the papers below:

- An agent based environmental impact assessment of building demolition waste management: Conventional versus green management. https://doi.org/10.1016/j.jclepro. 2016.06.054
- Cost-benefit analysis of demolition waste management via agent-based modelling: A case study in Shenzhen. https://doi.org/10.1016/j.wasman.2021.10.036
- An agent-based model approach for urban demolition waste quantification and a management framework for stakeholders. https://doi.org/10.1016/j.jclepro.2020.124897

3.5 Social Network Analysis

Social networks for analysis can be divided into one-mode networks and two-mode networks. The one-mode network only investigates networks with one set of nodes. Two-mode networks have two different sets of nodes. Additionally, links only exist between nodes that belong to different sets.

3.5.1 One-mode Social Network Analysis

Social network analysis concentrates on network relationships: interpersonal relationships, group relationships, organizational relationships, national relationships. A complete social network usually contains cohesion and brokerage. Through the analysis of these characteristics, key network nodes and network relations in the network can be found.

1) Cohesion

Network node cohesion often occurs in social networks, which indicates that some network nodes are closely related and some network nodes are alienated or not connected. Cohesion analysis is to explore the node with cohesion effects and the patterns and reason for the cohesion.

Density – Density is one of the most commonly used concepts in network analysis; it is used to represent the cohesion of the nodes in the network. Intuitively, more ties between the nodes yield a tighter network structure, which is, presumably, more cohesive (De Nooy et al. 2018). Density refers to the actual number of connections in a simple network, represented by the proportion of the largest possible number of connections in the network. If every two nodes in a network are connected to each other, then the network is called complete network. In the complete network, the connection ability of any two nodes is very strong. Even if we cut off part of the connection or remove some nodes, the two nodes can still connect through other paths.

The relationship between nodes can be directed or undirected. A directed relationship is represented by an arc while an undirected relationship is represented by an edge. Accordingly, the social network can be classified into a directed network and an undirected network. An undirected network contains no arcs, which means all relationships are edge. According to whether the connection is assigned, the network can be divided into weighted network and unweighted network. The connection in an assignment network is endowed with numerical value.

The range of density values of an assignment-free network is between 0 and 1. If all the nodes in a network do not have any connections, the density of the network is 0. If a network is a complete network, the density of the network is 1. The value range of the assignment network is not fixed, which depends mainly on the assignment.

The density of a directed network without assignment can be attained using the formula:

$$D = \frac{L}{2C_N^2} \tag{3.1}$$

where D represents density of network, L represents the number of edge, and N represent the number nodes.

The density of an undirected network without assignment can be attained using the formula:

$$D = \frac{L}{C_N^2} \tag{3.2}$$

The density of a directed network with assignment can be attained using the formula:

$$D = \frac{\sum L_w}{2C_N^2} \tag{3.3}$$

where $\sum L_w$ indicates the sum of all the assignments of lines in a network.

Degree – Network density is often restricted by network scale. The larger the social network, the lower the network density. The reason is that as the number of fixed points increases, the maximum number of possible connections will increase rapidly. Therefore, it is not enough to use only density when comparing different networks, degree also should be used. Node degree indicates the number of connections owned by a node in a social network. The higher the degree, the more likely the node appears in the high-density area. The average node degree is an important index used to measure the network cohesion. It is not affected by the network size and, therefore, can be used for comparison between different scale networks. In a directed network, degree can be categorized into in-degree and out-degree. The in-degree represents the number of edges directed into a vertex in a directed network; the out-degree represents the number of edges directed out of a vertex in a directed network. Degree difference indicates the difference between out-degree and in-degree. A node with a high degree difference will have stronger influences on its neighbours than it will accepting influence (Li et al. 2016). When a social network is with assignment, the calculation of node degree needs to consider the assignment of network. The calculation formulas of in-degree, out-degree, and degree difference are as:

$$InDegree_{S_I} = \sum_{S_{II} \in G} RSM_{S_{II}, S_I} \tag{3.4}$$

$$OutDegree_{S_I} = \sum_{S_{II} \in G} RSM_{S_I, S_{II}} \tag{3.5}$$

$$GapDegree_{S_I} = OutDegree_{S_I} - InDegree_{S_I} \tag{3.6}$$

where S_{II} refers to the nodes directly related to S_I, G refers to social network, RSM refers to network relation matrix, $InDegree_{S_I}$ refers to the in-degree of S_I, $OutDegree_{S_I}$ refers to the out-degree of S_I, and $GapDegree_{S_I}$ refers to the degree difference of S_I

2) Brokerage

Social network structure analysis helps to explain how information, goods, attitudes, or even behaviour diffuses within a social system (De Nooy et al. 2018). De Nooy et al. (2018) pointed out that 'some social structures permit rapid diffusion of information, whereas others contain sections that are difficult to reach'. The study of brokerage is to explore the key relationships and node in a social network. Brokerage includes centrality, centralization, and a brokerage role.

Centrality and centralization – centrality refers to positions of nodes within the network while centralization refers to the characteristics of an entire network. In a highly centralized network, information can be spread easily and timely.

The degree centrality of a node is node degree. A node with high out-degree will have strong influences on its neighbours while a node with high in-degree can be easily influenced by its neighbours. Degree centralization of a social network is the variation in the degree of nodes divided by the maximum degree variation as follows:

$$DC = \frac{\sum_{S_I \in G} (MaxDegree - Degree_{S_I})}{(N-1)(N-2)} \tag{3.7}$$

where DC is degree centralization, *MaxDegree* is the max degree of a node, G is the network and N is the number of nodes.

In an undirected network, the distance between two nodes is the number of edges in the shortest path (also called geodesic) connecting the two nodes. Different from undirected networks, the distance between two nodes may have two values since the directed network is a system of 'one-way street' (De Nooy et al. 2018).

The closeness centrality of a node is based on the total distance between one node to all other nodes (Wang 2015), which can be attained as follows:

$$CC(S_I) = \frac{n-1}{\sum_{S_{II} \in G} d(S_I, S_{II})} \tag{3.8}$$

where $CC(S_I)$ refers to the closeness centrality of S_I and G refers to the network.

Similar to degree centralization, closeness centralization is the variation in the closeness centrality of nodes divided by the maximum variation (De Nooy et al. 2018).

Degree and closeness centrality focus on the reachability of a node within a network. Betweenness is the percentage of times a node lies on the shortest path 'between' two other nodes (Wang 2015). Accordingly, the betweenness centrality of a node is the proportion of all geodesics between pairs of other nodes that include this node (De Nooy et al. 2018). Betweenness centrality of a node (S_I) can be attained as follows:

$$C_B(S_I) = \sum_{S_{II}} \sum_{S_{III}} \frac{g_{S_{II} S_{III}}(S_I)}{g_{S_{II} S_{III}}} \tag{3.9}$$

where $C_B(S_I)$ is the betweenness centrality of S_I, $g_{S_{II} S_{III}}(S_I)$ refers to the number of geodesics between S_{II}, and S_{III}, which includes S_I, $g_{S_{II} S_{III}}$, refers to the number of all geodesics between S_{II} and S_{III}.

Wang (2015) pointed out that 'the betweenness centrality is a measure for the control of information flows within the network and the function of single nodes as intermediaries'.

Brokerage roles – represent the role of the node in connecting different subgroups in a network. There are five kinds of brokerage roles, namely coordinator, itinerant broker, representative, gatekeeper, and liaison (Figure 3.17). In the first case, all the three nodes are in the same subgroups, *node 1* represents the coordinator between *node 2* and *node 3*. In the second case, two nodes of a subgroup use a mediator (itinerant broker) from outside. In the third case, *node 1* acts as a representative of the subgroup which regulates the flow of information; all the information from the subgroup to *node 3* should be transmitted through *node 1*. In the fourth case, *node 1* is a gatekeeper because it regulates the information from *node 2* to its subgroup. In the last case, node 1 acts as a liaison because it mediates between different members belong to another different subgroup.

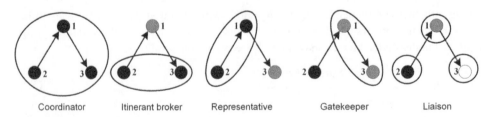

Coordinator   Itinerant broker   Representative   Gatekeeper   Liaison

Figure 3.17 Five brokerage roles in a social network.

3.5.2 Two-mode Social Network Analysis

Although most networks are considered as one-mode networks, some networks, also known as affiliation or bipartite networks, are in fact two-mode networks (Borgatti and Everett 1997; Latapy et al. 2008). An example of two-mode social network is present in Figure 3.18.

1) Degree centrality

Degree centrality focus on the reachability of a node within a network. It is an indicator of immediate popularity or connectivity of a node as well as its vulnerability to catching whatever flows though the network (Sankar et al. 2015). The simplest definition of centrality is that nodes in the centre should be the most active in a sense (Wasserman and Faust 1994). Compared with other nodes in the network, nodes in the centre should have the most relationships. Degree centrality can be calculated by the formula:

$$C_D(S_I) = \frac{\sum_{S_{II} \in G} RSM_{S_{II}, S_I}}{N} \tag{3.10}$$

where S_{II} refers to the nodes directly related to S_I, G refers to social network, RSM refers to network relation matrix, and N refers to the total number of nodes in the set that S_{II} belongs to.

A node with a high level of degree centrality often affects the behaviour of the network, while a node with a low level of degree centrality is located at the periphery of the network and has little impact on the network. If a node is completely isolated, the removal of this node from the network has no effect on the entire network (Wasserman and Faust 1994).

2) Betweenness centrality

Betweenness is the percentage of times a node lies on the shortest path 'between' two other nodes (Wang 2015). Accordingly, betweenness centrality of a node is the proportion of all geodesics between pairs of other nodes that include this node (De Nooy et al. 2018). Betweenness centrality of a node (S_I) can be attained by Formula (3.11):

Figure 3.18 An example of two-mode social network.

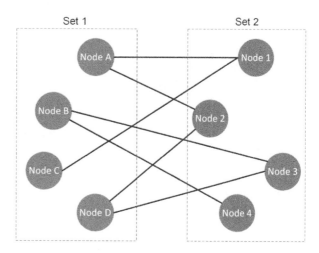

$$C_B(S_I) = \sum_{S_{II}} \sum_{S_{III}} \frac{g_{S_{II}S_{III}}(S_I)}{g_{S_{II}S_{III}}} \tag{3.11}$$

where $C_B(S_I)$ is the betweenness centrality of S_I, $g_{S_{II}S_{III}}(S_I)$ refers to the number of geodesics between S_{II}, and S_{III}, which includes S_I, $g_{S_{II}S_{III}}$, refers to the number of all geodesics between S_{II} and S_{III}.

Wang (2015) pointed out that 'the betweenness centrality is a measure for the control of information flows within the network and the function of single stakeholders as intermediaries'. It can be found that a node can be more powerful and influential if it can mediate and control more directed interactions between other nodes.

3) Eigenvector centrality

Eigenvector centrality, also known as eigencentrality, is a metric of the influence of a node in its network (Sankar et al. 2015). It measures how well connected are the nodes to which a node is connected (Sankar et al. 2015; Xue et al. 2018). It is based on the belief that a node is important if its neighbours are important. According to Borgatti and Everett (1997), the eigenvector centrality of a node can be calculated by Formula (3.12):

$$C_e = \sqrt{\frac{1}{2n_0}} \tag{3.12}$$

where n_0 is the size of the vertex set of the node belongs to.

4) Categorical core/periphery

Core/periphery structures commonly exist in social networks. They are usually decomposed into a dense cohesive core plus an outlying, loosely connected periphery (Zhang et al. 2015). One of the characteristics of core/periphery structures is that nodes in the core position cannot be divided into independent cohesive subgroups. Meanwhile, nodes in the periphery position are only closely related to some core nodes relative to each other, while the peripheral nodes are sparsely connected with each other and present scattered edge distribution. According to Borgatti and Everett (2000) and Xue et al. (2018), the core/periphery structure can be explored by Formula (3.13) and Formula (3.14):

$$\rho = \sum_{i,j} a_{ij} \delta_{ij} \tag{3.13}$$

$$\delta_{ij} = \begin{cases} 1 \text{ if } c_i = CORE \text{ or } c_j = CORE \\ 0 \text{ otherwise} \end{cases} \tag{3.14}$$

where a_{ij} represents the presence or absence of a link in the observed data, c_i indicates the core or periphery position that node i is assigned to, and δ_{ij} refers to the presence or absence of a link in the ideal structure.

3.5.3 Application Example

To demonstrate the process of how to use the social network analysis method, we use an example about 'Stakeholder collaboration and building energy performance gap: A two-mode social network analysis'. Energy-related stakeholders play a critical role in energy-saving and emission reduction. However, most previous studies focused on only a few stakeholders (e.g., occupant, designer, and owner). Therefore, there is a lack of systematic analysis of energy-related stakeholders. Improving building energy efficiency requires the collaboration of all stakeholders in the building lifecycle. This example applies a two-mode social network model to investigate the stakeholders' collaboration and its impact on building energy efficiency.

1) Identification of issues and stakeholders

A total of 16 impact issues of building energy performance gap are identified on the basis of the literature review and the semi-structured interviews as follows: lack of responsibility(I1), poor supervision (I2), standard and technology (I3), collaboration and communication (I4), variation (I5), limited cost and time (I6), improper design (I7), lack of information integrity (I8), poor quality of building and equipment (I9), incomplete commissioning (I10), porridge building operation (I11), indoor environment (I12), poor knowledge and experience (I13), rebound effect (I14), occupant comfort (I15), occupant behaviour (I16).

A total of 12 energy-related stakeholders are identified as follows: owner (S1), designer (S2), contractor (S3), subcontractor (S4), supervision (S5), manufacturer (S6), commissioning agent (S7), energy manager (S8), occupant (S9), policymakers and government agencies (S10), media (S11), researcher (S12).

2) Stakeholder–issue adjacency matrix

The developed adjacency matrix of the stakeholder–impact issue network is shown in Table 3.6. The horizontal dimension represents energy-related stakeholders and the vertical dimension refers to impact issues. If impact issue α can be addressed by stakeholder β, then the (α, β) entry in the adjacency matrix is 1; otherwise, then (α, β) entry is 0.

The issue–issue matrix can be obtained by converting the adjacency matrix. For example, four stakeholders, namely S2 (designer), S6 (manufacturer), S7 (commissioning agent), and S8 (energy manager), can address I3 (standard and technology) and I11 (porridge building operation) simultaneously. Thus, the (3,11) and (11,3) entry in the issue–issue matrix becomes 4 (Table 3.7).

The stakeholder–stakeholder matrix indicates the number of issues that each pair of stakeholder groups can address (Table 3.8). If stakeholder α and stakeholder β can address n issues simultaneously, then the (α, β) and (β, α) entry in the adjacency matrix is n.

3) Visualizing the stakeholder–issue network

The stakeholder–issue network is composed of 12 stakeholders and 16 impact issue nodes connected by 85 links. In Figure 3.19, the blue circles represent the stakeholders and red squares represent the impact issues. The links between stakeholders and issues represent the power of stakeholders over issues.

Along with the stakeholder–issue network, three indicators, namely degree centrality, betweenness centrality and, eigenvector centrality, are calculated. In Table 3.9, S2 (designer) has the highest degree centrality because it has power over all impact issues. S7 (commissioning agent), which addressed 11 issues, tied at second place. This finding suggests that these stakeholders have more power to bridge the building energy performance gap given that they can address most issues. Meanwhile, they have the potential to collaborate with a wider range of stakeholders. In terms of betweenness centrality, S2 (designer), S12 (researcher), and S7 (commissioning agent) have the highest scores, indicating that they have more control over the network.

Table 3.6 Adjacency matrix of the stakeholder–impact issue network.

	S1	S2	S3	S4	S5	S6	S7	S8	S9	S10	S11	S12	SUM
I1	1	1	1	1	1	1	1	1	0	1	1	1	11
I2	1	1	0	0	1	0	0	0	0	1	1	1	6
I3	0	1	1	1	1	1	1	1	0	1	0	1	9
I4	1	1	1	1	1	1	1	1	0	0	0	1	9
I5	1	1	1	1	0	0	0	0	0	0	0	0	4
I6	1	1	1	1	0	0	0	0	0	0	0	0	4
I7	0	1	0	0	0	0	0	0	0	0	0	1	2
I8	1	1	1	1	1	1	1	1	1	0	0	1	10
I9	0	1	1	1	1	1	1	0	0	0	0	0	6
I10	1	1	1	1	1	1	1	0	0	0	0	0	7
I11	1	1	0	0	0	1	1	1	1	0	0	0	6
I12	0	1	0	0	0	1	1	1	1	0	0	0	5
I13	1	1	1	1	1	1	1	1	1	0	0	1	10
I14	0	1	0	0	0	0	0	0	1	0	1	1	4
I15	0	1	0	0	0	0	1	1	0	0	0	1	4
I16	0	1	0	0	0	0	1	1	1	1	1	1	7
SUM	9	16	9	9	8	9	11	9	6	4	4	10	

Table 3.7 Adjacency matrix of the issue–issue matrix.

	I1	I2	I3	I4	I5	I6	I7	I8	I9	I10	I11	I12	I13	I14	I15	I16
I1	11	6	9	9	4	4	2	9	6	7	5	4	9	3	4	6
I2	6	6	4	4	2	2	2	4	2	3	2	1	4	3	2	4
I3	9	4	9	8	3	3	2	8	6	6	4	4	8	2	4	5
I4	9	4	8	9	4	4	2	9	6	7	5	4	9	2	4	4
I5	4	2	3	4	4	4	1	4	3	4	2	1	4	1	1	1
I6	4	2	3	4	4	4	1	4	3	4	2	1	4	1	1	1
I7	2	2	2	2	1	1	2	2	1	1	1	1	2	2	2	2
I8	9	4	8	9	4	4	2	10	6	7	6	5	10	3	4	5
I9	6	2	6	6	3	3	1	6	6	6	3	3	6	1	2	2
I10	7	3	6	7	4	4	1	7	6	7	4	3	7	1	2	2
I11	5	2	4	5	2	2	1	6	3	4	6	5	6	2	3	4
I12	4	1	4	4	1	1	1	5	3	3	5	5	5	2	3	4
I13	9	4	8	9	4	4	2	10	6	7	6	5	10	3	4	5
I14	3	3	2	2	1	1	2	3	1	1	2	2	3	4	2	4
I15	4	2	4	4	1	1	2	4	2	2	3	3	4	2	4	4
I16	6	4	5	4	1	1	2	5	2	2	4	4	5	4	4	5

Table 3.8 Stakeholder–stakeholder matrix.

	S1	S2	S3	S4	S5	S6	S7	S8	S9	S10	S11	S12
S1	9	9	7	7	6	6	6	5	3	2	2	5
S2	9	16	9	9	8	9	11	9	6	9	4	10
S3	7	9	9	9	7	7	7	5	2	2	1	5
S4	7	9	9	9	7	7	7	5	2	2	1	5
S5	6	8	7	7	8	7	7	5	5	4	2	6
S6	6	9	7	7	7	9	9	7	4	2	1	5
S7	6	11	7	7	7	9	11	9	5	3	2	7
S8	5	9	5	5	5	7	9	9	5	3	2	7
S9	3	6	2	2	5	4	5	5	6	1	2	3
S10	2	9	2	2	4	2	3	3	1	4	3	3
S11	2	4	1	1	2	1	2	2	2	3	4	3
S12	5	10	5	5	6	5	7	7	3	3	3	10

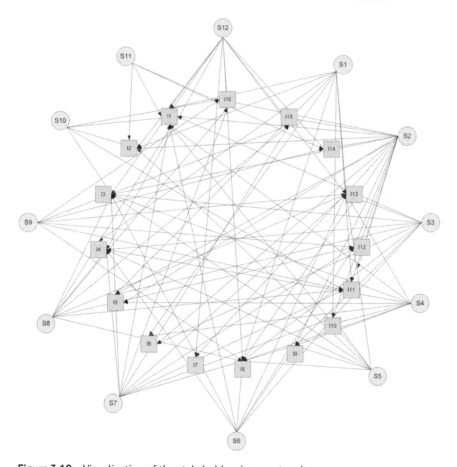

Figure 3.19 Visualization of the stakeholder–issue network.

Table 3.9 Centrality of stakeholders in the stakeholder–impact issue network.

Stakeholders	Degree centrality	Rank	Betweenness centrality	Rank	Eigenvector centrality	Rank
S1	0.563	4	0.046	4	0.284	9
S2	1.000	1	0.226	1	0.441	1
S3	0.563	4	0.036	6	0.303	4
S4	0.563	4	0.036	6	0.303	4
S5	0.500	9	0.025	9	0.287	8
S6	0.563	4	0.033	8	0.314	3
S7	0.688	2	0.062	3	0.358	2
S8	0.563	4	0.039	5	0.298	6
S9	0.375	10	0.020	10	0.172	10
S10	0.250	11	0.006	12	0.127	11
S11	0.250	11	0.009	11	0.104	12
S12	0.625	3	0.074	2	0.296	7

The issue with the highest degree centrality is I1 (responsibility), followed by I8 (information integrity) and I13 (knowledge and experience) (Table 3.10). All stakeholders must pay attention to these three issues because they cannot be handled by only one stakeholder. However, most stakeholders lack systematic thinking. This indicates that overcoming the building energy performance gap requires a change in the mindset of all stakeholders. The top three building energy performance gap issues with the highest eigenvector centrality are I1 (lack of responsibility), I8 (lack of information integrity), and I13 (poor knowledge and experience), which are the same as for degree centrality and betweenness centrality.

To demonstrate the process of how to use social network analysis method, please read the papers below:

- Stakeholder-associated risks and their interactions in complex green building projects: A social network model. https://doi.org/10.1016/j.buildenv.2013.12.014
- Modelling stakeholder-associated risk networks in green building projects. https://doi.org/10.1016/j.ijproman.2015.09.010
- Schedule risks in prefabrication housing production in Hong Kong: a social network analysis. https://doi.org/10.1016/j.jclepro.2016.02.123

3.6 Interpretive Structural Modelling

Interpretive structural modelling (ISM) identifies relationships among multiple influencing factors in complex socioeconomic systems (Shi et al. 2016; Yu et al. 2018). It is acknowledged that factors may be interrelated in a complex system. The direct and indirect relationships among factors may describe complex systems more reasonably and accurately

Table 3.10 Centrality of impact issues in the stakeholder–impact issue network.

Issues	Degree centrality	Rank	Betweenness centrality	Rank	Eigenvector centrality	Rank
I1	0.917	1	0.097	1	0.367	1
I2	0.500	8	0.031	7	0.181	11
I3	0.750	4	0.049	4	0.321	5
I4	0.750	4	0.038	6	0.340	4
I5	0.333	12	0.005	13	0.157	13
I6	0.333	12	0.005	14	0.157	13
I7	0.167	16	0.001	16	0.087	16
I8	0.833	2	0.057	2	0.360	2
I9	0.500	8	0.013	10	0.236	7
I10	0.583	6	0.019	8	0.270	6
I11	0.500	8	0.018	9	0.220	8
I12	0.417	11	0.011	12	0.186	10
I13	0.833	2	0.057	2	0.360	2
I14	0.333	12	0.013	10	0.119	15
I15	0.333	12	0.005	14	0.164	12
I16	0.583	6	0.042	5	0.212	9

than the individual factors taken into isolation (Attri et al. 2013). Therefore, there is a need for a method that can analyse multiple factors based on a system perspective (Shen et al. 2016). ISM is such a research method in that enables insight into both direct and indirect interrelationships among various factors with direction and ranking (Tan et al. 2019).

Compared with other alternative methods such as social network analysis (SNA), analytic hierarchy process (AHP), and decision-making trial and evaluation laboratory (DEMATEL), ISM has been chosen as an effective method when investigating relationships among various factors of the complex system (Abuzeinab et al. 2017; AL-Muftah et al. 2018; Luthra et al. 2014). ISM can not only provide insight into the interrelationships among different factors but also assist in finding the hierarchical way that factors are organized (Janes 1988). Moreover, the purpose behind using ISM is to impose order and direction on the complexity of relationships among elements of a system (Jindal and Sangwan 2011). Most importantly, 'ISM is an interactive learning process in which a set of dissimilar and directly related elements are structured into a comprehensive systematic model' (Mathiyazhagan et al. 2013).

ISM has been applied in the field of sustainable construction, e.g., analysing the execution of emission trading systems in the building sector (Shen et al. 2016), probing barriers to the transition towards off-site constriction (Gan et al. 2018), investigating critical factors for implementing sustainable construction practice (Yu et al. 2018), and developing a hierarchy of actions required to achieve the objective of waste management (Sharma and Gupta 1995).

3.6.1 ISM Development

The ISM method establishes the hierarchical structure between representative factors. The Adjacency Matrix is first developed to quantify the relationships among the factors. Since the Adjacency Matrix only shows the direct relationships among factors, and it cannot present the indirect relationships, it is necessary to conduct power iteration analysis (Gan et al. 2018). The Reachability Matrix is developed through power iteration analysis so that it can show both direct and indirect relationships (Gan et al. 2018). According to (Gan et al. 2018) and (Shen et al. 2016), the Reachability Matrix can be obtained through the following calculation process:

$$R = A^k, \text{when } A = A^{K=1} \tag{3.15}$$

where R is the Reachability Matrix, A is the adjacency matrix, and k is the number of iterations.

Once the reachability matrix is obtained, partitioning can be conducted to establish the hierarchy structure. The reachability set, antecedent set, and intersection set are the significant elements in partitioning. The reachability set of a factor contains all the factors (including itself) that it may reach, and the antecedent set of a factor consists of all the factors (including itself) that may reach to it (Gan et al. 2018; Shen et al. 2016). The intersection set is the intersection of the reachability set and the antecedent set. Factors with the same intersection set and reachability set are considered as pertaining to level *I*. Subsequently, the Level *I* factors will be discarded for the next iteration.

3.6.2 MICMAC

MICMAC is coupled with ISM, in other words, these two methods must be applied in combination. MICMAC aims to investigate the diffusion of impacts 'through reaction paths and loops for developing hierarchies for members of an element set' (Wang et al. 2008). It is an analysis method to classify factors into four clusters according to the driving power and dependence power of each factor, which is always integrated with ISM (Shen et al. 2016; Yu et al. 2018). Walters et al. (2018) opined that MICMAC is a structural factor analysis method to explore indirect relationships and feedback loops among factors. Luthra et al. (2014) and Wang et al. (2008) proved that MICMAC is a useful method to analyse the dependence power and driving power of each factor in order to have a clear understanding of the interactive relationship among them. The driving power of a factor refers to the total number of other factors affected by it, while the dependence power indicates the total number of other factors affecting it. The data input of MICMAC is the data output of ISM (Shi et al. 2016). Based on MICMAC, all the factors can be classified into four categories:

1) Driving factors: these factors are strong in driving power while weak in dependence power. They are more capable of influencing other factors. Therefore, more attention should be paid on driving factors.
2) Linkage factors: these factors have strong driving power and dependence power. They are sensitive in nature. Any action on these factor will have a reaction effect on the other factors as well as on themselves (Ali et al. 2018; Shen et al. 2016).

3) Autonomous factors: these factors have weak driving power and dependence power. They have few links to the system they belong to, which means that they do not easily affect other factors in the system, nor are they easily influenced by other factors.

4) Dependent factors: these factors are weak in driving power and strong in dependence power. Dependent factors are usefully influenced by linkage factors and driving factors, and they are less likely to affect other factors. Therefore, if driving factors and linkage factors are addressed, dependent factors will be addressed accordingly (Shen et al. 2016).

3.6.3 Application Example

The following section provides an example on how to use ISM method to identify the factors and their complex relationships analysis on the building energy performance gap. This example is drawn from research previously conducted by the Xu and Zou (2020). For more details about the research and publication, readers are directed to read https://doi.org/10.1016/j.jclepro.2020.122650.

The gap between the designed and measured building energy consumption is one of the biggest obstacles to the realization of the energy conservation goal. The factor analysis method has been typically used in previous studies to explore the individual factors affecting the building energy performance gap, while few studies examined the complex interrelationships among the influencing factors. In this research, the ISM approach was developed and applied to identify the key factors and explore the interrelationships among the factors affecting the building energy performance gap.

1) Representative factors identification

A list of 16 representative factors that affect the building energy performance gap are identified based on literature review and expert interview. They are rationality of design (F1), knowledge and experience (F2), information integrity (F3), responsibility (F4), collaboration and communication (F5), variation (F6), supervision (F7), standard and technology (F8), cost and time (F9), quality of building and equipment (F10), commissioning (F11), building operation (F12), and indoor environment (F13), rebound effect (F14), occupant comfort (F15), and occupant behaviour (F16).

2) Building Adjacency Matrix

The contextual relationships among 16 representative factors are structured in an Adjacency Matrix based on the feedback from the 12 experts (Table 3.11). It can be found that F6 (cost and time) directly affects most of other factors while no factor could directly influence F6.

3) Developing Reachability Matrix

The Reachability Matrix (Table 3.12) is developed by applying Formula (3.15). It is based on the principle that if Factor α can affect Factor β and Factor β can affect Factor γ, then Factor α can necessarily affect Factor γ.

4) Establishing the hierarchy structure

The reachability set, antecedent set, and intersection set can be identified based on the Reachability Matrix. Different levels can be found by comparing the elements of each factor of the reachability set and intersection set. It can be found that Level *I* contains factor

Table 3.11 The Adjacency Matrix (A) between 16 representative factors.

A	F1	F2	F3	F4	F5	F6	F7	F8	F9	F10	F11	F12	F13	F14	F15	F16
F1	0	0	0	1	1	0	1	1	1	1	1	0	0	0	0	0
F2	1	0	0	1	0	0	1	1	1	1	1	0	0	0	0	0
F3	0	0	0	0	0	0	0	0	1	0	0	0	1	0	0	0
F4	0	0	0	0	1	0	1	1	1	1	1	0	1	0	0	0
F5	0	0	0	0	0	0	0	1	1	0	0	0	0	0	0	0
F6	1	0	1	1	1	0	1	1	1	1	1	0	0	0	0	0
F7	0	0	0	0	1	0	0	0	1	1	1	1	0	0	1	1
F8	0	0	0	0	1	0	1	0	0	1	1	0	1	0	0	0
F9	0	0	0	0	0	0	0	0	0	0	1	1	0	0	0	0
F10	0	0	0	0	0	0	0	0	0	0	1	1	0	0	0	0
F11	0	0	0	0	0	0	0	0	0	0	0	1	0	0	0	0
F12	0	0	0	0	0	0	0	0	0	0	0	0	0	0	1	0
F13	0	0	0	1	1	0	1	1	1	1	1	0	0	0	0	0
F14	0	0	0	0	0	0	0	0	0	1	0	0	0	0	0	1
F15	0	0	0	0	0	0	0	0	0	0	0	0	0	0	0	1
F16	0	0	0	0	0	0	0	0	0	0	0	1	0	0	0	0

Table 3.12 Reachability Matrix (R) between 16 representative factors.

A	F1	F2	F3	F4	F5	F6	F7	F8	F9	F10	F11	F12	F13	F14	F15	F16
F1	1	0	0	1	1	0	1	1	1	1	1	1	1	0	1	1
F2	1	1	0	1	1	0	1	1	1	1	1	1	1	0	1	1
F3	0	0	1	1	1	0	1	1	1	1	1	1	1	0	1	1
F4	0	0	0	1	1	0	1	1	1	1	1	1	1	0	1	1
F5	0	0	0	1	1	0	01	1	1	1	1	1	1	0	1	1
F6	1	0	1	1	1	1	1	1	1	1	1	1	1	0	1	1
F7	0	0	0	1	1	0	1	1	1	1	1	1	1	0	1	1
F8	0	0	0	1	1	0	1	1	1	1	1	1	1	0	1	1
F9	0	0	0	0	0	0	0	0	1	0	1	1	0	0	1	1
F10	0	0	0	0	0	0	0	0	0	1	1	1	0	0	1	1
F11	0	0	0	0	0	0	0	0	0	0	1	1	0	0	1	1
F12	0	0	0	0	0	0	0	0	0	0	0	1	0	0	1	1
F13	0	0	0	1	1	0	1	1	1	1	1	1	1	0	1	1
F14	0	0	0	0	0	0	0	0	0	0	0	1	0	1	1	1
F15	0	0	0	0	0	0	0	0	0	0	0	1	0	0	1	1
F16	0	0	0	0	0	0	0	0	0	0	0	1	0	0	1	1

12 (indoor environment), 15 (occupant comfort), and 16 (occupant behaviour) as they have the same intersection set and reachability set (Table 3.13).

In Table 3.14, factors 12, 15, and 16 as well as the rows corresponding to factors 12, 15, and 16 are eliminated from Table 3.13. Factors 11 (building operation) and 14 (rebound effect) are put at Level *II* as the reachability and intersection set for them are the same. The same procedure of eliminating rows corresponding to previous level and identifying the next level location in the new table is repeated until the final level is identified.

It is inferred that Level *III* includes two factors, namely Factors 9 (quality of building and equipment) and 10 (commissioning) as they have the same intersection set and reachability set (Table 3.15).

In Table 3.16, the elements of factors 4 (collaboration and communication), 5 (variation), 7 (rationality of design), 8 (information integrity), and 13 (knowledge and experience) at reachability set and intersection set are the same. Therefore, they are placed at Level *IV*.

In Table 3.17, Factors 1 (responsibility) and 3 (standard and technology) are found to have the same intersection set and reachability set. Therefore, they are put at Level *V*.

Finally, in Table 3.18 factors 2 (supervision) and 6 (cost and time) are placed at the highest level, suggesting their higher influences on the building energy performance gap and they are likely to affect other factors.

Based on the analysis result from Tables 3.13–3.18, the ISM based hierarchy structure among 16 representative factors is attained, as shown in Figure 3.20.

Table 3.13 Level 1 partition.

Factor	Reachability set	Antecedent set	Intersection set	Level
1	1,4,5,7,8,9,10,11,12,13,15,16	1,2,6	1	
2	1,2,4,5,7,8,9,10,11,12,13,15,16	2	2	
3	3,4,5,7,8,9,10,11,12,13,15,16	3,6	3	
4	4,5,7,8,9,10,11,12,13,15,16	1,2,3,4,5,6,7,8,13	4,5,7,8,13	
5	4,5,7,8,9,10,11,12,13,15,16	1,2,3,4,5,6,7,8,13	4,5,7,8,13	
6	1,3,4,5,6,7,8,9,10,11,12,13,15,16	6	6	
7	4,5,7,8,9,10,11,12,13,15,16	1,2,3,4,5,6,7,8,13	4,5,7,8,13	
8	4,5,7,8,9,10,11,12,13,15,16	1,2,3,4,5,6,7,8,13	4,5,7,8,13	
9	9,11,12,15,16	1,2,3,4,5,6,7,8,9,13	9	
10	10,11,12,15,16	1,2,3,4,5,6,7,8,10,13	10	
11	11,12,15,16	1,2,3,4,5,6,7,8,9,10,11,13	11	
12	12,15,16	1,2,3,4,5,6,7,8,9,10,11,12, 13,14,15,16	12,15,16	*I*
13	4,5,7,8,9,10,11,12,13,15,16	1,2,3,4,5,6,7,8,13	4,5,7,8	
14	12,14,15,16	14	14	
15	12,15,16	1,2,3,4,5,6,7,8,9,10,11,12, 13,14,15,16	12,15,16	*I*
16	12,15,16	1,2,3,4,5,6,7,8,9,10,11,12, 13,14,15,16	12,15,16	*I*

Table 3.14 Level 2 partition.

Factor	Reachability set	Antecedent set	Intersection set	Level
1	1,4,5,7,8,9,10,11,13	1,2,6	1	
2	1,2,4,5,7,8,9,10,11,13	2	2	
3	3,4,5,7,8,9,10,11,13	3,6	3	
4	4,5,7,8,9,10,11,13	1,2,3,4,5,6,7,8,13	4,5,7,8,13	
5	4,5,7,8,9,10,11,13	1,2,3,4,5,6,7,8,13	4,5,7,8,13	
6	1,3,4,5,6,7,8,9,10,11,13	6	6	
7	4,5,7,8,9,10,11,13	1,2,3,4,5,6,7,8,13	4,5,7,8,13	
8	4,5,7,8,9,10,11,13	1,2,3,4,5,6,7,8,13	4,5,7,8,13	
9	9,11	1,2,3,4,5,6,7,8,9,13	9	
10	10,11	1,2,3,4,5,6,7,8,10,13	10	
11	11	1,2,3,4,5,6,7,8,9,10,11,13	11	*II*
13	4,5,7,8,9,10,11,13	1,2,3,4,5,6,7,8,13	4,5,7,8	
14	14	14	14	*II*

Table 3.15 Level 3 partition.

Factor	Reachability set	Antecedent set	Intersection set	Level
1	1,4,5,7,8,9,10,13	1,2,6	1	
2	1,2,4,5,7,8,9,10,13	2	2	
3	3,4,5,7,8,9,10,13	3,6	3	
4	4,5,7,8,9,10,13	1,2,3,4,5,6,7,8,13	4,5,7,8,13	
5	4,5,7,8,9,10,13	1,2,3,4,5,6,7,8,13	4,5,7,8,13	
6	1,3,4,5,6,7,8,9,10,13	6	6	
7	4,5,7,8,9,10,13	1,2,3,4,5,6,7,8,13	4,5,7,8,13	
8	4,5,7,8,9,10,13	1,2,3,4,5,6,7,8,13	4,5,7,8,13	
9	9	1,2,3,4,5,6,7,8,9,13	9	*III*
10	10	1,2,3,4,5,6,7,8,10,13	10	*III*
13	4,5,7,8,9,10,13	1,2,3,4,5,6,7,8,13	4,5,7,8,13	

Table 3.16 Level 4 partition.

Factor	Reachability set	Antecedent set	Intersection set	Level
1	1,4,5,7,8,13	1,2,6	1	
2	1,2,4,5,7,8,13	2	2	
3	3,4,5,7,8,13	3,6	3	
4	4,5,7,8,13	1,2,3,4,5,6,7,8,13	4,5,7,8,13	*IV*
5	4,5,7,8,13	1,2,3,4,5,6,7,8,13	4,5,7,8,13	*IV*
6	1,3,4,5,6,7,8,13	6	6	
7	4,5,7,8,13	1,2,3,4,5,6,7,8,13	4,5,7,8,13	*IV*
8	4,5,7,8,13	1,2,3,4,5,6,7,8,13	4,5,7,8,13	*IV*
13	4,5,7,8,13	1,2,3,4,5,6,7,8,13	4,5,7,8	*IV*

Table 3.17 Level 5 partition.

Factor	Reachability set	Antecedent set	Intersection set	Level
1	1	1,2,6	1	V
2	1,2	2	2	
3	3	3,6	3	V
6	1,3,6	6	6	

Table 3.18 Level 6 partition.

Factor	Reachability set	Antecedent set	Intersection set	Level
2	2	2	2	VI
6	6	6	6	VI

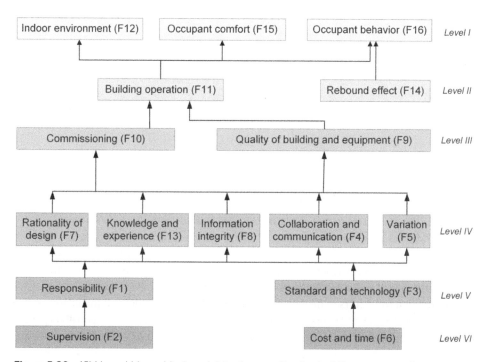

Figure 3.20 ISM based hierarchical model for factors affecting building energy performance gap.

5) Driving power and dependence power of representative factors

Factors affecting the building energy performance gap are grouped into four categories based on their driving power and dependence power, as shown in Table 3.19 and Figure 3.21. Figure 3.21 comprises four quadrants, representing driving factors, linkage factors, autonomous factors, and dependent factors respectively. For example, factor 1 (responsibility) with

Table 3.19 Driving power and dependence power of each factor.

R	F1	F2	F3	F4	F5	F6	F7	F8	F9	F10	F11	F12	F13	F14	F15	F16
Driving-power	12	13	12	11	11	14	11	11	5	5	4	3	11	4	3	3
Dependence-power	3	1	2	9	9	1	9	9	10	10	12	16	9	1	16	16

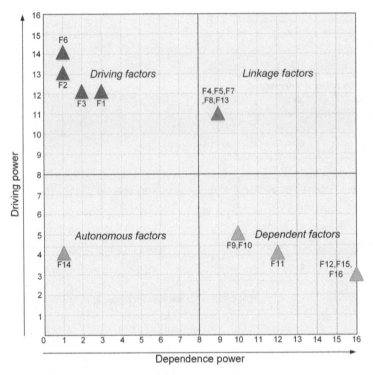

Figure 3.21 The driving power and dependence power of factors affecting building energy performance gap.

a dependence power of 3 and a driving power of 12 is placed at a position of 3 on the *x*-axis and 12 on the *y*-axis, and it is defined as a driving factor

It is found that the factors of the bottom levels are all driving factors and the factors of the top levels are all dependent factors. 'Supervision (F2)', 'cost and time (F6)', 'responsibility (F1)', and 'standard and technology (F3)' can be considered as root causes of the building energy performance gap, as they have the strongest capability to affect other factors and are less likely to be influenced by other factors. 'Indoor environment (F12)', 'occupant comfort (F15)', and 'occupant behaviour (F16)' can be seen as direct causes of the building energy performance gap, as they cannot affect other factors.

The following papers provide examples on how to use the ISM method to identify the influencing factors and their complex relationships:

- Interpretive Structural Modelling based factor analysis on the implementation of Emission Trading System in the Chinese building sector. https://doi.org/10.1016/j.jclepro.2016.03.151

- Barriers to Building Information Modelling (BIM) implementation in China's prefabricated construction: An interpretive structural modelling (ISM) approach. https://doi.org/10.1016/j.jclepro.2019.02.141

3.7 Data Mining Methods

Data mining as a method for data analysis has been applied in many areas of research in recent years. While the following sections provide brief discussions, Chapter 8 provides more details.

3.7.1 Statistical Analysis

Statistical analysis is a science of collecting, exploring, and presenting large amounts of data to uncover patterns and trends (Sprinthall and Fisk 1990). Descriptive statistics analysis, probability distribution analysis, and linear regression analysis are usually applied. Descriptive statistics describes the basic feature of the data. Probability distribution analysis is a method that allows us to create a statistical distribution with a specific pattern to describe the probability; from probability distribution, we can estimate the probability when an event might be likely to occur. Linear regression is applied to analyse variation over time in the proportion of events. All data analysis can be performed using R, a free software for statistical computing and graphics.

3.7.2 Data Visualization

Data visualization refers to the techniques used to present data by encoding it as visual objects in a graphic format. It is a central part of exploratory data analysis (Myatt and Johnson 2009). Compared with traditional descriptive statistics, data visualization can distil large datasets into graphics to allow for easy understanding of trends and complex relationships within the data.

3.7.3 Association Rule Mining

Association rule mining (ARM) is an unsupervised learning method to discover hidden relationships and correlations in large datasets (Xiao and Fan 2014). It was first introduced for discovering regularities between products in supermarkets (Agrawal et al. 1993). Currently, ARM has been applied in the field of sociology (Hastie et al. 2009) and building energy (Fan et al. 2015). The rules are based on the frequency number of an item set in combination with other sets in a database (Verma et al. 2014). Brief descriptions of the ARM method are provided in the following sections.

Let α be a large item set, an associate rule is defined as an implication of the form $X \rightarrow Y$, where $X, Y \subset \alpha$ and $X \cap Y = \varnothing$. The set X is called the antecedent and the set Y is called the consequent. There are two measurements of rule effectiveness, i.e. support and confidence (Verma et al. 2014). Support reflects the usefulness of discovered rules, and it is the joint probability of the antecedent and consequent (Formula (3.16)). Confidence refers to certainty of discovered rules and it is an estimate of the conditional probability, as shown in Formula (3.17). Only those rules that meet the thresholds of support and confidence are considered to be meaningful (Fan et al. 2015). The thresholds of support and confidence are usually set to 0.006 and 0.25, respectively.

$$\text{Support}(X \rightarrow Y) = P(X, Y) \tag{3.16}$$

$$\text{Confidence}(X \rightarrow Y) = P(Y|X) \tag{3.17}$$

Lift is a measure of 'interestingness' of a discovered rule. It is defined as the correlation between the occurrence of X and Y (Formula (3.18)). Specifically, a lift value equal to 1 means that the occurrence of X and Y are independent of each other. A lift value greater than 1 indicates the occurrence of X and Y are dependent on each other, while a lift value smaller than 1 means the occurrence of X and Y are substitute to each other.

$$\text{Lift}(A \rightarrow B) = (P(X,Y))/(P(X)P(Y)) = (P(Y|X))/(P(Y)) \tag{3.18}$$

More detail about how to apply data science in research will be discussed in Chapter 7.

3.8 Criticism about Quantitative Research

Similar to qualitative research, there are criticisms of quantitative research that are worth outlining. Quantitative research fails to distinguish people and social institutions from the natural world, and thus ignores the fact that people interpret the world around them. This capacity that cannot be found among the objects of the natural sciences, such as molecules, cells, and materials (Bryman, 2016). Some researchers argue that quantitative research is value-free, i.e., objective, but in truth no one can be fully detached from any type of research because the researchers themselves influence and shape their research based on certain assumptions about the world, through the accumulated knowledge that they have learnt (Grix 2004). As such, bias may occur in quantitative research as the actual behaviour of respondents may differ from their answers (Bryman, 2016).

3.9 Summary

Quantitative research is one of the most frequently used research methods. In this chapter, instead of following the classical processes and steps, for which there is a wide amount of existing literature, we have focused on hypothesis development, design, and testing as well as structural equation modelling. We then presented the concepts and techniques of the commonly used quantitative analytical methods, together with worked practical applications examples. These include system dynamics, agent-based modelling, social network analysis, interpretive structural modelling, and several data mining methods. Data mining methods are also related to Chapter 7, Data-Driven Research. Like the previous chapter on qualitative research, there is criticism about quantitative research which is mentioned in this chapter. To overcome the shortcomings of qualitative and quantitative research, the next chapter will discuss mixed methods research, where, as the name implies, both qualitative research and quantitative research are 'mixed' to provide a more complete set of methods to explain the complex phenomena or answer complex research problems and achieve research aims.

Review Questions and Exercises

1 How do you develop a hypothesis in research?
2 How do you test hypothesis in research?
3 What is your understanding of 'system dynamics' as a quantitative method in research?
4 What are the main steps in using system dynamics analytical methods?
5 Describe the steps in social network analysis.
6 What are the specific steps for using the interpretive structural modelling (ISM) method?
7 What is your understanding of data mining? How can this method be used in research?

References

Abuzeinab, A., Arif, M., and Qadri, M.A. (2017). Barriers to MNEs green business models in the UK construction sector: an ISM analysis. *Journal of Cleaner Production* 160: 27–37.

Agrawal, R., Imielinski, T., and Swami, A. (1993). Mining association rules between sets ofitems in large databases, *Proceedings of the 1993 ACM SIGMOD InternationalConference on Management of Data* 207–216.

Ahmad, S., Tahar, R.M., Muhammad-Sukki, F. et al. (2016). Application of system dynamics approach in electricity sector modelling: a review. *Renewable and Sustainable Energy Reviews* 56: 29–37.

Akhwanzada, S.A. and Tahar, R.M. (2012). Strategic forecasting of electricity demand using system dynamics approach. *International Journal of Environmental Science and Development* 3 (4): 328.

Ali, S.M., Arafin, A., Moktadir, M.A. et al. (2018). Barriers to reverse logistics in the computer supply chain using interpretive structural model. *Global Journal of Flexible Systems Management* 19: 53–68.

Al-Muftah, H., Weerakkody, V., Rana, N.P. et al. (2018). Factors influencing e-diplomacy implementation: exploring causal relationships using interpretive structural modelling. *Government Information Quarterly* 35 (3): 502–514.

Arbuckle, J. and Wothke, W. (2004). Structural equation modeling using AMOS: an introduction.

Attri, R., Dev, N., and Sharma, V. (2013). Interpretive structural modelling (ISM) approach: an overview. *Research Journal of Management Sciences* 2 (2): 3–8.

Babbie, E.R. (2020). *The Practice of Social Research*, 15e. Cengage Learning.

Balci, O. (1995). Validation, verification, and testing techniques throughout the life cycle of a simulation study. *Annals of Operations Research* 53 (1): 121–173.

Barlas, Y. (1996). Formal aspects of model validity and validation in system dynamics. *System Dynamics Review* 12 (3): 183–210.

Borgatti, S.P. and Everett, M.G. (1997). Network analysis of 2-mode data. *Social Networks* 19: 243–269.

Borgatti, S.P. and Everett, M.G. (2000). Models of core/periphery structures. *Social Networks* 21: 375–395.

Borshchev, A. (2013) Multi-method modeling. In: *Proceedings of the 2013 Winter Simulation Conference* (eds S.H. Kim, A. Tolk, R. Hill, et al.), 4089–4100.

Bryman, A. (2016). *Social Research Methods*. Oxford; New York: Oxford University Press.

Byrne, B.M. (2013). *Structural Equation Modeling with Mplus: Basic Concepts, Applications, and Programming*. Routledge.

De Nooy, W., Mrvar, A., and Batagelj, V. (2018). *Exploratory Social Network Analysis with Pajek*. Cambridge University Press.

Ding, Z., Hu, T., Li, M. et al. (2019). Agent-based model for simulating building energy management in student residences. *Energy and Buildings* 198: 11–27.

Fan, C., Xiao, F., and Yan, C. (2015). A framework for knowledge discovery in massivebuilding automation data and its application in building diagnostics. *Automation in Construction* 50: 81–90.

Gan, X., Chang, R., Zuo, J. et al. (2018). Barriers to the transition towards off-site construction In China: an interpretive structural modeling approach. *Journal of Cleaner Production* 197: 8–18.

Grix, J. (2004). *The Foundation of Research*. Hampshire: Palgrave Macmillan.

Hastie, T., Tibshirani, R., and Friedman, J. (2009). *The Elements of Statistical Learning: Data Mining, Inference, and Prediction*. Springer Science & Business Media.

Iba, H. (2013). *Agent-based Modeling and Simulation with Swarm*. CRC Press.

Janes, F. (1988). Interpretive structural modelling: a methodology for structuring complex issues. *Transactions of the Institute of Measurement and Control* 10: 145–154.

Jin, X., Xu, X., Xiang, X. et al. (2016). System-dynamic analysis on socio-economic impacts of land consolidation in china. *Habitat International* 56: 166–175.

Jindal, A. and Sangwan, K.S. (2011) Development of an interpretive structural model of barriers to reverse logistics implementation in Indian industry. Glocalized solutions for sustainability in manufacturing. Springer.

Latapy, M., Magnien, C., and Del Vecchio, N. (2008). Basic notions for the analysis of large two-mode networks. *Social Networks* 30: 31–48.

Li, C.Z., Hong, J., Xue, F. et al. (2016). Schedule risks in prefabrication housing production in Hong Kong: a social network analysis. *Journal of Cleaner Production* 134: 482–494.

Li, Z., Shen, G.Q., and Alshawi, M. (2014). Measuring the impact of prefabrication on construction waste reduction: an empirical study in china. *Resources, Conservation and Recycling* 91: 27–39.

Luthra, S., Kumar, S., Kharb, R. et al. (2014). Adoption of smart grid technologies: an analysis of interactions among barriers. *Renewable and Sustainable Energy Reviews* 33: 554–565.

Mathiyazhagan, K., Govindan, K., Noorulhaq, A., and Geng, Y. (2013). An ISM approach for the barrier analysis in implementing green supply chain management. *Journal of Cleaner Production* 47: 283–297.

Mohamed, S. (2002). Safety climate in construction site environments. *Journal of Construction Engineering and Management* 128 (5): 375–384.

Myatt, G.J. and Johnson, W.P. (2009). *Making Sense of Data II: A Practical Guide to Datavisualization, Advanced Data Mining Methods, and Applications*. John Wiley.

Ogunlana, S.O., Li, H., and Sukhera, F.A. (2003). System dynamics approach to exploring performance enhancement in a construction organization. *Journal of Construction Engineering and Management* 129 (5): 528–536.

Russell, S. and Norvig, P. (1995). A modern, agent-oriented approach to introductory artificial intelligence. *ACM Sigart Bulletin* 6 (2): 24–26.

Sankar, C.P., Asokan, K., and Kumar, K.S. (2015). Exploratory social network analysis of affiliation networks of Indian listed companies. *Social Networks* 43: 113–120.

Senge, P.M. and Forrester, J.W. (1980). Tests for building confidence in system dynamics models. *TIMS Studies in the Management Sciences* 14: 209–228.

Sharma, H. and Gupta, A. (1995). The objectives of waste management in India: a futures inquiry. *Technological Forecasting and Social Change* 48: 285–309.

Shen, L., Song, X., Wu, Y. et al. (2016). Interpretive structural modeling based factor analysis on the implementation of emission trading system in the Chinese building sector. *Journal of Cleaner Production* 127: 214–227.

Shi, Q., Yu, T., Zuo, J., and Lai, X. (2016). Challenges of developing sustainable neighborhoods in China. *Journal of Cleaner Production* 135: 972–983.

Sprinthall, R.C. and Fisk, S.T. (1990). *Basic Statistical Analysis*. EnglewoodCliffs, NJ: Prentice Hall.

Sterman, J.D. (2000). *Business Dynamics: Systems Thinking and Modeling for a Complex World.* McGraw-Hill Education.

Taeger, D. and Kuhnt, S. (2014). *Statistical Hypothesis Testing with SAS and R*. John Wiley.

Tan, T., Chen, K., Xue, F., and Lu, W. (2019). Barriers to building information modeling (BIM) implementation in China's prefabricated construction: an interpretive structural modeling (ISM) approach. *Journal of Cleaner Production* 219: 949–959.

Taylor, S.J.E. (2014). *Agent-based Modeling and Simulation*. Springer.

Verma, A., Khan, S.D., Maiti, J., and Krishna, O.B. (2014). Identifying patterns of safety related incidents in a steel plant using association rule mining of incident investigation reports. *Safety Science* 70: 89–98.

Walters, J., Kaminsky, J., and Gottschamer, L. (2018). A systems analysis of factors influencing household solar PV adoption in Santiago, Chile. *Sustainability* 10: 1257.

Wang, G., Wang, Y., and Zhao, T. (2008). Analysis of interactions among the barriers to energy saving in China. *Energy Policy* 36: 1879–1889.

Wang, H. (2015). *The Interplay of Formal and Informal Institutions for Procurement Innovation: A Social Network Approach*. Doctor of Philosophy, The University of Hong Kong.

Wasserman, S. and Faust, K. (1994). *Social Network Analysis: Methods and Applications*. Cambridge University Press.

Wilensky, U. and Rand, W. (2015). *An Introduction to Agent-based Modeling: Modeling Natural, Social, and Engineered Complex Systems with NetLogo*. MIT Press.

Xiao, F. and Fan, C. (2014). Data mining in building automation system for improving building operational performance. *Energy and Buildings* 75: 109–118.

Xu, X.X., Wang, J.Y., Li, C.Z.D. et al. (2018). Schedule risk analysis of infrastructure projects: a hybrid dynamic approach. *Automation in Construction* 95: 20–34.

Xu, X.X. and Zou, P.X.W. (2020). Analysis of factors and their hierarchical relationships influencing building energy performance using interpretive structural modelling (ISM) approach. *Journal of Cleaner Production* 272: 122650.

Xu, X.X. and Zou, P.X.W. (2021). System dynamics analytical modeling approach for construction project management research: a critical review and future directions. *Frontiers of Engineering Management* 8 (1): 17–31.

Xue, X., Zhang, X., Wang, L. et al. (2018). Analyzing collaborative relationships among industrialized construction technology innovation organizations: a combined SNA and SEM approach. *Journal of Cleaner Production* 173: 265–277.

Yu, T., Shi, Q., Zuo, J., and Chen, R. (2018). Critical factors for implementing sustainable construction practice in HOPSCA projects: a case study in China. *Sustainable Cities and Society* 37: 93–103.

Yuan, H. and Wang, J. (2014). A system dynamics model for determining the waste disposal charging fee in construction. *European Journal of Operational Research* 237 (3): 988–996.

Zhang, X., Martin, T., and Newman, M.E. (2015). Identification of core-periphery structure in networks. *Physical Review E* 91: 032803.

Zou, P.X.W. and Sunindijo, R.Y. (2013). Skills for managing safety risk, implementing safety task, and developing positive safety climate in construction project. *Automation in Construction* 34: 92–100.

4

Mixed Methods Research

4.1 Introduction

Chapters 1, 2, and 3 discussed different research paradigms, specifically qualitative and quantitative research methods. Considering the strengths and weaknesses of different methods, it would be naturally logical to mix the qualitative and quantitative methods. The nature of most research nowadays is a combination of social science and natural science, which indicates that a mixed methods would be useful. This chapter provides an in-depth and comprehensive discussion of mixed methods research. A mixed methods research framework is developed for researchers to gain a comprehensive understanding and implementation of mixed methods research design, according to the nature of their research needs, research problems, and research aims.

4.2 Comparing Qualitative Research and Quantitative Research

Quantitative research is a theoretical *test*, while qualitative research is a theoretical *construction*. Qualitative research starts from empirical facts, establishes the relationship between facts, and establishes theories through the analysis of data; Quantitative research starts from theory and tests the theory with empirical data. While qualitative research is generally based on constructivism and interpretivism, quantitative research is based on positivism and postpositivism. From an ontology perspective, qualitative research emphasizes individuality and uniqueness and adheres to subjective theory construction. On the contrary, quantitative research recognizes the causality of the objective world and believes that the objective world exists independently of human perception. Chapter 1 has presented substantial discussion on this topic.

Due to different philosophical standpoints, there are differences in the applicability of qualitative research and quantitative research. Qualitative research is mainly applicable to exploratory, interpretative, and complex problems. The purpose of qualitative research tends to clarify concepts or construct new theories to provide specific guidance for practice. Quantitative research is mainly applicable to descriptive, relational, and causal questions. The goal of quantitative research is to collect facts to test theory, measure some aspect of a phenomenon or trend, and identify causality.

Research Methodology and Strategy: Theory and Practice, First Edition. Patrick X.W. Zou and Xiaoxiao Xu.
© 2023 John Wiley & Sons Ltd. Published 2023 by John Wiley & Sons Ltd.

4.3 Definition and Current Applications of Mixed Methods Research

'Mixed methods research is the type of research in which a researcher or team of researchers combines elements of qualitative and quantitative research approaches (e.g., qualitative and quantitative viewpoints, data collection, analysis, inference techniques) for the broad purposes of breadth and depth of understanding and corroboration' (Johnson et al. 2007). Problems observed and experienced tend to be varied, uncertain, and complex, and thus difficult to address (Carayon et al. 2015). This highlights the need for mixed methods research design. Mixed methods research can help enrich and improve the understanding of phenomena that cannot be fully understood by using single methodology (Lopez-Fernandez and Molina-Azorin 2011; Venkatesh et al. 2013). One of the main reasons for using mixed methods research is the need for multiple data sources and data collection and analysis methods to address research problems that are too complex to be approached one-dimensionally. In short, mixed methods research is becoming increasingly articulated and recognized as the third research methodology, along with qualitative research and quantitative research (Clark et al. 2023).

Although some journals have traditionally been oriented toward qualitative research, such as *Sociologia Ruralis* and *Rural Studies* (Strijker et al. 2020), or quantitative research, such as *Rural Sociology* (Strijker et al. 2020) and *Tourism Management* (Hewlett and Brown 2018), in recent years, more attention has been paid to the design and application of mixed methods research across fields (Hewlett and Brown 2018; Strijker et al. 2020). In fact, mixed methods research has been increasingly applied in many fields, such as education (Clark et al. 2023), sustainability (Bhutta et al. 2021), rural studies (Strijker et al. 2020), energy management (Zou et al. 2018), tourism management (Hewlett and Brown 2018), health care (Carayon et al. 2015), supply chain management (Dubey et al. 2015), and safety management (Zou et al. 2014). These applications of mixed methods research are explained below.

- Zou et al. (2014) suggested that a greater use of mixed methods research will better integrate the realms of theory and practice. They developed a mixed methods research design that incorporates several different research methods, inductively and deductively, and encourages iteration between the realms of theory and practice. This mixed methods research design demands the combination of multiple forms of research in order to provide insights to 'what', 'why', and 'how' types of research questions.
- Dubey et al. (2015) highlighted the use of mixed methods research to address issues related to green supply chain management. Interpretive structured modelling, MICMAC analysis, and confirmatory factor analysis were used to illustrate the application of mixed methods research by testing a model on the green supply chain management enablers.
- Carayon et al. (2015) reviewed the application of mixed methods research in human factors and ergonomics research. They observed an increasing number of mixed methods research in health care. In particular, they found that the way for mixing qualitative research and quantitative research includes convergent parallel design, sequential design, embedded design, and multiphase design. They also found that a variety of methods were used for collecting data, but mainly including interview for qualitative data, survey for quantitative data, and observation for both.

- Zou et al. (2018) designed a mixed methods framework for building occupants' energy consumption behaviour research. They provided four scenarios: (i) a need exists to contrast or compare qualitative and quantitative findings or to validate or expand quantitative results with qualitative data; (ii) a need exists to include qualitative (or quantitative) data within a largely quantitative (or qualitative) study to solve a research problem; (iii) a need exists to explain the quantitative results; and (iv) a need exists to first explore qualitatively.

- Hewlett and Brown (2018) discussed the use of mixed methods research to address contemporary issues and challenges in the tourism industry. Using a unique mixed methods research, they demonstrated the value of both qualitative and quantitative data to tourism planners, and the ability of mixed methods research through geographical information system to visualize a highly subjective and value-laden concept.

- Strijker et al. (2020) discussed the pros and cons of mixed methods research in rural studies and concluded that the rural research context offers considerable scope for the application of mixed methods research. Specifically, since economics and parts of sociology are more quantitatively oriented, and geography and other parts of sociology are more qualitatively oriented, rural studies sit at the crossroads of both qualitative research and quantitative research.

- By using mixed methods research, Bhutta et al. (2021) investigated complex relationships among environmental sustainability, innovation capacity, and stakeholders in the context of sustainable supply chain management. The mixed methods research used in this context consists of designing a survey instrument, collecting data from a sample of organizations, developing a structural equation model, and conducting a cluster analysis.

- Clark et al. (2023) distinguished different types of frameworks for conceptualizing mixed methods research and illustrated how to apply a socioecological framework for mixed methods research within educational research practice.

4.4 Design of Mixed Methods Research

Carayon et al. (2015) specified four types of mixed methods research design:

1) Convergent parallel design: both qualitative and quantitative data are collected simultaneously and then combined at the stages of analysis and interpretation.
2) Sequential design: either qualitative or quantitative data collected first, then followed by the collection of the other type of data.
3) Embedded design: qualitative research (or quantitative research) is embedded in quantitative research (or qualitative research.
4) Multiphase design: combine concurrent and sequential collection and analysis of quantitative and qualitative data over a period of time.

Venkatesh et al. (2013) and Creswell and Clark (2017) also specified four types of mixed methods research design, with slightly different names: exploratory design, explanatory design, embedded design, and triangulation design. In this text, we adapt the latter four and the following sections provide discussions about each of these designs.

4.4.1 Exploratory Design

At the beginning of a research project, researchers may not be aware of which questions need to be proposed, which variables need to be measured and which theories can be based on, expanded, or revised. This lack of understanding may be due to the specificity of the research subject or the novelty of the research question. In such cases, it is best to first understand which questions, variables, and theories need to be studied through qualitative research, and then conduct quantitative research to verify the qualitative findings. The results of qualitative research can help develop or inform quantitative research when measures or instruments are not available, the variables are unknown, or there is no guiding framework or theory (Creswell and Clark 2017). Exploratory design is a two-stage design. Particularly in areas or directions where there is no large prior research base, the researcher first uses qualitative analysis to gain an initial understanding of the research question and to aid subsequent quantitative research (Greene et al. 1989). Researchers using exploratory design build on the results of the qualitative phase to develop an instrument that identifies variables or states propositions for testing based on an emerging theory or framework (Creswell and Clark 2017). These developments link the initial qualitative phase with the subsequent quantitative part of the research; because the design begins with the qualitative, there is usually more emphasis on qualitative data.

4.4.2 Explanatory Design

Sometimes quantitative research alone does not provide a complete explanation of research findings. In such cases, the researcher may use explanatory design. For example, the statistical significance, confidence intervals, and effect values generated by the results of quantitative studies provide generalized results for research. However, when we get such results, we are often unaware of the reasons and contextual scenarios that led to the results. Under such situations, qualitative data and results can help in understanding the significance of quantitative results.

Explanatory design is the use of qualitative data to explain or build upon initial quantitative results (Creswell 2021). Explanatory design is well suited to researchers who need to use qualitative data to explain uncommon quantitative findings, such as unexpected results, outliers or differences between groups need further exploration (Morse 1991). The main feature of an explanatory design is first a quantitative analysis followed by a qualitative analysis, with the aim of explaining or reinforcing the quantitative results (Creswell 2021).

Explanatory design consists of two stages: quantitative data collection and analysis, and qualitative data collection and analysis based on the quantitative results of the first stage. As this design starts with quantitative research, the researcher places more emphasis on quantitative methods than qualitative methods.

4.4.3 Embedded Design

Embedded design uses quantitative data to provide a supportive, secondary role in a study based primarily on qualitative data, or vice versa (Creswell 2021). In some cases, researchers need to answer questions involving different types of data, when it is clear that a single dataset will not meet the needs of the research. We recommend that researchers use an embedded design when they need to include qualitative data in primarily quantitative research to answer a research question, and vice versa.

Embedded design mixes different datasets, where one type of data is embedded in a methodology consisting of another data type (Caracelli and Greene 1997). Embedded design includes both qualitative and quantitative data collection, with one type of data serving as the main body of research and the other playing a complementary role. The inclusion of a secondary data type in a complementary role in a design based on the other data types distinguishes embedded design from other methods.

4.4.4 Triangulation Design

Triangulation design is about the use of qualitative research to corroborate quantitative research findings or vice versa. The purpose of triangulation design is 'to obtain different but complementary data on the same topic' (Morse 1991). In conducting research, qualitative data can provide detail, while quantitative data provide more holistic information. Specifically, qualitative data generally come from studies of individuals and in-depth exploration of their perspectives, while quantitative data come primarily from examining large populations and determining their perspectives on multiple variables. However, one type of data may not tell the complete story, or there may be a need to consider more than one type of data to sufficiently answer the research question. There may also be contradictions in the evidence from qualitative and quantitative data that would not be detected if only one type of data was collected (Zou et al. 2018).

In the above situations, researchers often use triangulation design to contrast and compare quantitative research results with qualitative findings or to validate or expand quantitative results with qualitative data (Creswell and Clark 2017; Zou et al. 2014). It is the most common and well-known approach in mixed methods. This design allows for a combination of the advantages of quantitative methods (e.g., large sample size, trends, generalization) with the advantages of qualitative methods (e.g., small sample size, specific, in-depth) (Patton 1990).

The triangulation design is a single-stage design. Specifically, the researcher implements quantitative and qualitative methods of equal weight over the same time period. It usually involves simultaneous but separate collection and analysis of quantitative and qualitative data. This process also helps the researcher best understand the research question and its implications. Researchers attempt to merge the two types of data, typically by putting separate results together in an interpretation or by transforming the data to facilitate the integration of the two data types in the analysis (Zou et al. 2018).

4.5   Framework of Mixed Methods Research

In many cases it is inevitable that both qualitative and quantitative data are used. There is a lack of mixed methods research framework for researchers to apply. In order to fill the research gap, a mixed methods research framework is proposed as illustrated in Figure 4.1. The proposed framework has the ability to combine different disciplines whilst emphasizing the importance of the research problem.

As shown in Figure 4.1, the research should commence from identifying the research problem (stage 1), since mixed methods research is research problem oriented. Researchers

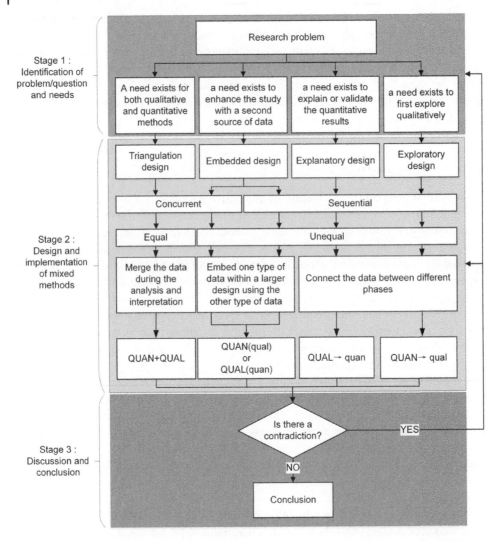

Figure 4.1 Mixed methods research framework (Zou et al. 2018 / with permission of Elsevier).
Note: QUAN refers to quantitative research with greater weight; QUAL refers to qualitative research with greater weight; quan refers to quantitative research with less weight; and qual refers to qualitative research with less weight.

need to turn a practical problem into theoretical problem. To solve this problem, research-ers should examine existing studies and theories related to the research area (Zou et al. 2014). In this stage, researchers should also make it clear what is needed for solving the identified problem.

There are four kinds of needs for conducting mixed methods research.

1) A need exists for both qualitative and quantitative methods. This is when only one type of research method cannot tell the complete story, or the researcher may lack

confidence in the ability of one type of research method to solve research problem (Creswell and Clark 2017). For instance, to explore the relationship between building occupants' requirements and their behaviour, Guerra-Santin et al. (2016) investigated indoor environment and building occupant behaviour from the quantitative point of view and occupants' attitudes toward energy saving from the qualitative point of view.

2) A need exists to enhance the study with a second source of data. Researchers need a qualitative design (or quantitative design) to enhance quantitative data (or qualitative data).

3) A need exists to explain or validate the quantitative results. This indicates that quantitative results need validation or interpretation with the help of qualitative methods. For example, interview or workshop could be used to explain the workers' unsafe behaviour that is uncovered through data mining.

4) A need exists to first explore qualitatively (Creswell and Clark 2017). Qualitative research should be conducted to identify variables, constructs, and theories prior to using quantitative research. For some research problems, there is no mature theory or framework for researchers to learn from. Thus, a qualitative method should be used first to provide an adequate exploration before a quantitative method is used to provide further understanding (Li et al. 2017).

If a researcher has one or more of the above four needs, they should consider using mixed methods research, and choose and implement a mixed methods research design according to their needs (Stage 2). The four needs correspond to exploratory design, explanatory design, embedded design, and triangulation design respectively (Figure 4.1). When selecting a mixed methods research design, researchers should ask and answer three questions (Creswell and Clark 2017).

1) What will the timing of the qualitative and quantitative methods be?
2) What will the relative weighting of the qualitative and quantitative methods be?
3) How can qualitative and quantitative methods be mixed?

Timing relates to when the data are collected, analysed, and interpreted. In Figure 4.1, symbol '+' means a simultaneous design, while '→' indicates a sequential design. Weighting refers to the relative importance of the qualitative and quantitative methods to answer identified problem (Creswell and Clark 2017). In the design framework, the method with greater weight is shown in capital letters (QUAL, QUAN), otherwise it is written in lower case (qual, quan). Mixing is the most important procedure in mixed methods design, as pointed out by Creswell and Clark (2017); research that contains qualitative and quantitative methods without mixing the data from each other is just a multimethod. For triangulation design, qualitative and quantitative methods should be conducted concurrently and equally, with both types of data merged during analysis and interpretation. In contrast, qualitative and quantitative methods in an explanatory or exploratory design should be sequential and unequal. Since qualitative and quantitative methods are applied in two phases in explanatory or exploratory designs, their data should be connected between the two phases. In embedded design, qualitative and quantitative methods can be concurrent or sequential but one type of data should be embedded within a larger design using the other type of data.

In Stage 3, researchers must be aware of any contradiction in the mixed methods design. For example, qualitative research cannot explain the results of quantitative research or the conclusion of the qualitative research in conflict with the conclusion of the quantitative research. Under these circumstances, researchers need to rethink the research problem or reconstruct the mixed methods research design. If no contradiction exists throughout the research process, researchers could draw a conclusion.

4.6 Example of Conducting Mixed Methods Research

Four examples of mixed methods design are provided with the consideration of characteristics of different research methods, as shown in Table 4.1.

4.6.1 Example of Exploratory Design

Grounded theory, content analysis, case study, or interview methods could be applied in *exploratory design* to propose a theory or framework. Modelling and simulation, structural equation modelling (SEM), or path analysis could be used to develop instrument or quantitatively generalize the qualitative results. For example, if a researcher attempts to study the impact of user experience on energy use behaviour; as there is a paucity of current research on user experience, the researcher could first explore qualitatively the factors influencing user experience and the mechanism of its effect on energy use behaviour. In this phase, the researcher may select 60 participants for semi-structured interviews. The interviews generate live interview transcripts that can then be transcribed into interview texts. Using a thematic analysis procedure, the researcher identifies, from the qualitative dataset, the factors influencing user embodiment and the possible pathways through which user experience influences energy use behaviour. Based on the qualitative results, the researcher can design scales to measure the influencing factors of user experience. Following this, the researcher can administer the questionnaire to receive sufficient number of respondents and use a structural equation modelling (SEM) method to determine the pathways of influence.

Table 4.1 Four types of mixed methods research design.

Design type	Notation	Example
Exploratory	QUAL → quan	Grounded theory, content analysis, case study or interview → Modelling and simulation, structural equation modelling or path analysis
Explanatory	QUAN → qual	Modelling and simulation, traditional statistics or data mining → Interview or workshop
Embedded	QUAN (qual)	Modelling and simulation (interview or workshop)
Triangulation	QUAN+QUAL	Traditional statistics + interview
	QUAL (quan)	Grounded theory, content analysis, case study or interview (traditional statistics)

4.6.2 Example of Explanatory Design

In *explanatory design*, the results of qualitative research through interview or workshop methods can be used to explain or validate the results of quantitative modelling and simulation, traditional statistics or data mining. For instance, the researcher may use agent-based modelling to investigate the impact of user behaviour on building energy consumption. The findings present a quantitative relationship between different types of user behaviours and building energy consumption. However, the researcher may find it was difficult to interpret the research results and findings and, therefore, may carry out several rounds of expert interviews to provide interpretations of the results and findings through collecting and analysing the experts' extensive experience.

4.6.3 Example of Embedded Design

Embedded design is usually dominated by one method and supplemented by the other, that is, the other method is included under the framework of one method, and the other method plays a complementary and confirmatory role. There are two forms in embedded design. In the first form, the qualitative data from interview or workshop can be embedded within modelling and simulation with the intent of explaining mechanism. For instance, the researcher may apply quantitative modelling and simulation to investigate the impact of risks on the construction schedule of a large infrastructure project. The main body of this research design is the quantitative method. However, the qualitative component of the research is required because the researcher was not clear about the influence mechanism between the risk factors. Therefore, the researcher may select and interview, say, 15 risk management experts to obtain the influence mechanism.

In the second form, traditional statistics can be embedded with qualitative methods such as grounded theory, content analysis, case study, or interview by compiling text statistics. For example, the researcher may collect interview data from a number of experts and apply grounded theory to explore the barriers hindering building energy retrofit projects. In analysing the data, the researcher may find that the use of text-based statistical analysis could help identify the frequency of different barriers, and thus embed textual statistical analysis into the grounded theory.

4.6.4 Example of Triangulation Design

The specific process of *triangulation design* includes the following steps: (i) collect and analyse quantitative and qualitative data on the same phenomenon separately; and (ii) look for commonalities, differences, inconsistencies, or relationships between the results of quantitative research and qualitative research. In triangulation design, quantitative traditional statistics methods and qualitative interview methods are used to tell a complete story or reduce the deficiency of using only one type of the methods. For example, a researcher may select a certain number (say 60) of the construction workers as the study sample (N=60). The researcher then creates a database and adds in the quantitative data and qualitative data collected from 60 individuals. For quantitative data, the researcher may collect data on indicators about construction companies' safety culture. The researcher may also collect data on demographic characteristics from each participating company. The 60 construction workers may be divided into different groups according to their type of work. In analysing the quantitative data, the researcher may conduct descriptive statistics and group comparisons to examine whether there was significant difference between different groups on the concerned variables. For the qualitative data, the researcher conducts semi-structured interviews to understand how workers perceive safety culture. After transcribing the qualitative data obtained from interviews, the researcher analyses the text using a strategy of constant comparison to generate themes. In this case the quantitative and qualitative analysis are independent of each other. The researcher uses a qualitative method to group the respondents according to their perspectives, and uses descriptive statistics to compare the similarity and differences between different groups. The two types of data complement each other and are equally important in the research.

4.7 Summary

As the name implies, mixed methods research brings together qualitative research and quantitative research. Mixed methods research is being increasingly applied in many disciplines. This chapter has discussed the key aspects and developed a framework of mixed methods research. It started with a comparison of qualitative research and quantitative research, then moved to defining what mixed methods research is as well as its current state of application. The chapter then discussed four types of mixed methods research design: exploratory design, explanatory design, embedded design, and triangulation design. A methodological framework has been developed and detailed examples provided to demonstrate the applications of each of these four designs.

While there are many reasons and strengths for using mixed methods research, it should not be concluded that mixed methods research is definitively more effective than qualitative research or quantitative research; it is necessary to realize the advantages and disadvantages of these methods and their application conditions. Some critics argue that qualitative and quantitative methods carry different epistemological commitments that cannot be combined, such as the idea that qualitative data is multidimensional while quantitative data is one-dimensional and fixed, which could cause the loss of information when converting qualitative data into quantitative data. Mixed methods research is based on problem-centred and real-world practice-orientated pragmatism. Overall, the choice of a particular research method or methods rests on the nature, aim, and scope of the research and the research problems and questions.

Review questions and exercises

1   What are the similarities and differences between qualitative research and quantitative research?
2   What does mixed methods research mean?
3   What are the advantages and disadvantages of mixed methods research?
4   When would it be more advantageous to use mixed methods research?

5 What are the types of mixed methods research design?

6 Is an hypothesis needed in mixed methods research?

7 How would you apply a mixed methods research design in your field of research?

References

Bhutta, M.K., Muzaffar, A., Egilmez, G. et al. (2021). Environmental sustainability, innovation capacity, and supply chain management practices nexus: a mixed methods research approach. *Sustainable Production and Consumption* 28: 1508–1521.

Caracelli, V.J. and Greene, J.C. (1997). Crafting mixed-method evaluation designs. *New Directions for Evaluation* 74: 19–32.

Carayon, P., Kianfar, S., Li, Y. et al. (2015). A systematic review of mixed methods research on human factors and ergonomics in health care. *Applied Ergonomics* 51: 291–321.

Clark, V.L.P., Ivankova, N.V., and Yang, N. (2023). Frameworks for conceptualizing mixed methods research. In: *International Encyclopedia of Education* 4e (eds. R. Tierney, F. Rizvi, and K. Ercikan), 390–401. Elsevier.

Creswell, J.W. (2021). *A Concise Introduction to Mixed Methods Research*, 2e. SAGE Publications.

Creswell, J.W. and Clark, V.L.P. (2017). *Designing and Conducting Mixed Methods Research*, 3e. SAGE publications.

Dubey, R., Gunasekaran, A., Papadopoulos, T., and Childe, S.J. (2015). Green supply chain management enablers: mixed methods research. *Sustainable Production and Consumption* 4: 72–88.

Greene, J.C., Caracelli, V.J., and Graham, W.F. (1989). Toward a conceptual framework for mixed-method evaluation designs. *Educational Evaluation and Policy Analysis* 11 (3): 255–274.

Guerra-Santin, O., Herrera, N.R., Cuerda, E., and Keyson, D. (2016). Mixed methods approach to determine occupants' behaviour – analysis of two case studies. *Energy and Buildings* 130: 546–566.

Hewlett, D. and Brown, L. (2018). Planning for tranquil spaces in rural destinations through mixed methods research. *Tourism Management* 67: 237–247.

Johnson, R.B., Onwuegbuzie, A.J., and Turner, L.A. (2007). Toward a definition of mixed methods research. *The Journal of Mixed Methods Research* 1 (2): 112e133.

Li, D., Menassa, C.C., and Karatas, A. (2017). Energy use behaviors in buildings: towards an integrated conceptual framework. *Energy Research and Social Science* 23: 97–112.

Lopez-Fernandez, O. and Molina-Azorin, J.F. (2011). The use of mixed methods research in the field of behavioural sciences. *Quality & Quantity* 45 (6): 1459.

Morse, J.M. (1991). Approaches to qualitative-quantitative methodological triangulation. *Nursing Research* 40 (2): 120–123.

Patton, M.Q. (1990). *Qualitative Evaluation and Research Methods*. SAGE Publications, inc.

Strijker, D., Bosworth, G., and Bouter, G. (2020). Research methods in rural studies: qualitative, quantitative and mixed methods. *Journal of Rural Studies* 78: 262–270.

Venkatesh, V., Brown, S.A., and Bala, H. (2013). Bridging the qualitative-quantitative divide: guidelines for conducting mixed methods research in information systems. *MIS Quarterly* 37 (1): 21–54.

Zou, P.X.W., Sunindijo, R.Y., and Dainty, A.R. (2014). A mixed methods research design for bridging the gap between research and practice in construction safety. *Safety Science* 70: 316–326.

Zou, P.X.W., Xu, X., Sanjayan, J., and Wang, J. (2018). A mixed methods design for building occupants' energy behavior research. *Energy and Buildings* 166: 239–249.

5

Case Study Research

5.1 Introduction

As a method of research, case study is a small-scale study that aims to produce valuable data and analysis to help understand social phenomenon. Case study is in fact as difficult as any research method and needs to follow a strict set of logic. This chapter discusses the definitions of, arguments for, and types of case study, the processes and methods for data collection and analysis, and theory development in the context of case study.

5.2 Definitions of Case Study

A 'case' is a spatially bounded phenomenon observed at a given time or over a period of time. A case study is an intensive study of a single case or multiple cases that promises to shed light on a larger population of cases (Gerring 2017). A case study is often an empirical inquiry that investigates a contemporary phenomenon in depth and within its real-world context, especially when the boundaries between phenomenon and context may not be clearly evident (Yin 2014).

5.3 Philosophical Paradigms of Case Study

There are three philosophical paradigms of case study from the epistemological standpoint of researchers: interpretative, constructivist, and positivist which are discussed in the following sections.

1) *Interpretative*: Involves understanding meanings and contexts and processes as perceived from different perspectives, trying to understand individuals and shared social meanings. Interpretivism allows researchers to have multiple views on a problem, as it allows researchers to see the world through the eyes of the participants (Crowe et al. 2011).

2) *Constructivist*: Accepts that people construct their own understanding and knowledge of the world, through experiencing things and reflecting on those experiences. The focus is on theory building (Crowe et al. 2011).

Research Methodology and Strategy: Theory and Practice, First Edition. Patrick X.W. Zou and Xiaoxiao Xu.
© 2023 John Wiley & Sons Ltd. Published 2023 by John Wiley & Sons Ltd.

3) *Positivist*: Adheres to the view that only 'factual' knowledge gained through observation, including measurement, is trustworthy. This involves establishing which variables to study and the focus is on testing and refining theory on the basis of case study findings (Crowe et al. 2011).

5.4 Types of Case Study Methods

Case study can be used to explore, describe, or explain phenomena in the everyday contexts in which they occur (Yin 2014). Table 5.1 provides descriptions of each of these types of case studies.

5.5 When to Use Case Study Methods

The most important thing in conducting a case study is to determine whether the case study method is appropriate to the subject under investigation. To determine the appropriateness

Table 5.1 Descriptions of different types of case studies.

Types	Descriptions
Exploratory case study	This type of case study aims at identifying a new hypothesis. The researchers begin with a factor that is presumed to have a fundamental influence on a range of outcomes. Commonly, researchers normally work backward from a known outcome to possible cause (Gerring 2017).
Descriptive case study	The descriptive case study aims to look at the background of a person or groups and use a descriptive narrative about how they address a real-world situation.
Explanatory case study	The goal is to test a hypothesis by estimating a causal effect, which may mean precise point estimate along with a confidence interval or a less precise estimate of the sign of a relationship, i.e., whether X has a positive, negative, or no relationship with Y.
Intrinsic case study	The focus is to understand the unique phenomenon that is critical to a particular case only. In intrinsic case study researchers restrain curiosity and special interest and try to discern and pursue issues critical to the case. The findings may not be transferable to other cases.
Instrumental case study	The focus is to gain a broader understanding of a phenomenon through a particular case study. Here the phenomenon is more important (Stake 1995) than the detailed understanding of the case. The findings are transferable to other cases.
Collective case study	This involves studying multiple case studies simultaneously or sequentially in an attempt to generate broader appreciation of a particular phenomenon and test the theory.
Longitudinal case study	The focus is to observe a phenomenon over a period of time, and to construct concepts and theories explaining how phenomena emerge, evolve and terminate over the period of that time frame.

of a case study method, the nature of the problem under study should be considered. Case studies are best suited to practical issues in which the experience of the subjects is central and the context of the experience is decisive. Case study is suitable for examining phenomenon in which the associated factors are not controllable. Researchers should define the research objective and consider the existing theoretical base of the research subject. This entails identifying the phenomenon researchers are seeking to understand, its context, and the main issues it raises.

Case study is just one type of research method. When selecting research methods, we need to consider the following questions, which are cited from Yin (2014):

1) What is the type of research problem?
2) Is the focus of research a contemporary phenomenon?
3) Are the factors associated with the phenomenon controllable?

Conditions and processes of selecting different research methods are presented in Figure 5.1. The first step is to identify the type of research problem. In general, research problems can be divided into two types. Firstly, 'Who', 'What', 'Where', and 'When' type; These are descriptive and exploratory research problems. Descriptive case study, exploratory case study, and

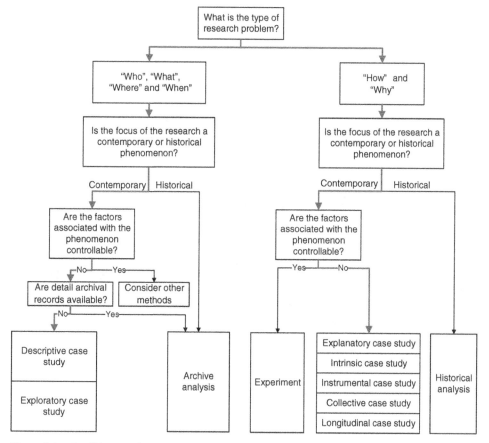

Figure 5.1 Conditions and processes for selecting different research methods.

archival analysis could be used to address this type of research problem. Secondly, 'How' and 'Why' type. This type of research problem is the explanatory research problem. Historical analysis, experiment, explanatory case study, intrinsic case study, instrumental case study, collective case study, and longitudinal case study may be suitable for addressing this type of research problem.

The second step is to clarify whether the focus of the research is a contemporary phenomenon and whether the factors associated with the phenomenon are controllable. If the research problem belongs to the 'How' and 'Why' type, and the phenomenon is historical, historical analysis is an appropriate method. When a contemporary phenomenon is not controllable by the researchers, case study is an appropriate method. This means that explanatory case study, intrinsic case study, instrumental case study, collective case study, and longitudinal case study are suitable for investigating phenomena that occur in contemporary times but the associated factors are *not controllable*. For example, health issues in elder people during the COVID-19 virus infection are a contemporary phenomena, but it would be unethical to conduct a controlled study that requires purposely infecting the older people with the coronavirus. Therefore, doctors can only rely on the anecdotal evidence provided by case studies to draw conclusions and make recommendations.

5.6 Processes of Implementing Case Study

5.6.1 Research Problem Identification

Researchers should carefully consider research problems and existing literature and theories. Formulating the right research problem is fundamental to conducting a successful case study. As the case study progresses, there may be a need to modify the research problem based on the observation and analysis of collected data (Yin 2017).

5.6.2 Research Hypothesis Development

Chapters 1 and 3 discussed the general process for hypothesis development. This section focuses on case study-based hypothesis development or testing. All research hypotheses should contain at least one independent variable and one dependent variable. Case study generally belongs to one of the three categories: (i) independent variable-centred case study, which focuses on the influencing factors of a specific cause; (ii) dependent variable-centred case study, which focuses on explaining a specific result; and (iii) independent–dependent variable-centred case study, which focuses on the relationships between specific causes and specific results. The first two case study categories can be regarded as exploratory research with the purpose of developing research hypotheses. The third case study category is related to confirmatory and falsifiable research, which aims to test hypotheses.

5.6.3 Purpose of Case Selection

The purpose of the case selection is to ensure the most suitable cases can be identified and selected prior to conducting data collection. Unless the case to be studied is unique, there may be a number of cases available. Generally speaking, the selection of cases depends on the type of case study that one has planned to do. For example, in an intrinsic case study, where the focus is to understand a unique case, the case is normally preselected. In the case of an instrumental case study, it may be useful to select cases that are typical or representative of other cases (Stake 1995). The following questions should be considered while selecting a case or cases.

- Does the case have the potential to help understand the research question and maximize the availability of valid information?
- Does it represent the research subjects?
- Do I have access to the case information?
- Is the time frame sufficient to conduct the case study?

5.6.4 Methods for Selecting Case

There are nine methods of case selection: typical case selection, diverse case selection, extreme case selection, deviant case selection, influential case selection, critical case selection, pathway case selection, most similar case selection, and most dissimilar case selection (Gerring 2017).

1) Typical case selection

A typical case is a representative of a wider range of cases and therefore contributes to the understanding of more general phenomena. Typicality refers to the mean, median, or plurality on a particular dimension. In most cases, the case selection criteria may be multiple dimensions. For instance, if we plan to select cities with the following characteristics for case study: (i) a long history, (ii) a high level of economic growth, and (iii) a distinctive local character, then Beijing, London, and Paris can be selected as typical cities. The characteristics in this example can be seen as criteria for case selection. Mathematically, the typicality of a case can be calculated using the formula:

$$\text{Typicality}(i) = -\left| y_i - \text{E}(y_i) \right| \tag{5.1}$$

$$\text{E}(y_i) = \beta_0 + \beta_1 x_{i,1} + \beta_2 x_{i,2} + \ldots + \beta_n x_{i,n} \tag{5.2}$$

where Typicality(i) refers to the typicality of Case i; y_i refers to value of the dependent variable of Case i; $x_{i,n}$ refers to the values of n independent variables of Case i; $\text{E}(y_i)$ refers to the expected value of y_i; and β_n refers to regression coefficient of $x_{i,n}$.

The closer the typicality of a case is to zero, the higher the typicality of the case. If researchers want to conduct a single-case study, they should select the case with the highest typicality. If researchers want to conduct a multiple-case study, they should select a group of cases with high typicality.

2) Diverse case selection

The aim of diverse case selection is to obtain maximum coverage in the concerned variables. If the research is only concerned with one independent variable which is classified, it is necessary to select cases from every classification. For example, if we want to investigate undergraduate students' perception of their majors and grade level is the single variable, we need to select typical cases from each grade level. For continuous variables, it is better for researchers to select cases representing extreme value, mean value, or median value. For instance, if a researcher plans to explore the attractiveness of cities to talent, and the research is focused on the GDP of cities which is a continuous variable, the researchers should select cities with the highest GDP, the lowest GDP, and the average GDP as cases to study.

If multiple variables need to be considered and each variable is a classification variable, the intersection between classifications needs to be considered. Researchers need to select one case from each intersection set. Specifically, researchers could select cases according to the formula:

$$N = \alpha_1 \alpha_2 \ldots \alpha_k \alpha_{k+1} \ldots \alpha_n \tag{5.3}$$

where N refers to the total number of intersections for all categories; α_k refers to the number of categories of the kth variable; and n refers to the total number of variables.

If variables are continuous, researchers first need to convert the variable into a classified variable according to its numerical value and then follow the above process. For instance, age, as a continuous variable, can be converted into two categories: one category is greater than 30 years old and the other category is less than or equal to 30 years old.

3) Extreme case selection

Extreme case refers to the case with extreme values in the independent or dependent variables. For example, If the variables in most cases have negative values, the ones with positive values are extreme cases, and vice versa (Seawright and Gerring 2008).

4) Deviant case selection

Deviant case refers to the case with anomalous values in the concerned variables. The deviation degree can be used to evaluate the anomalous degree of the case, as below:

$$\text{Deviation degree}(i) = |y_i - E(y_i)| \tag{5.4}$$

$$E(y_i) = \beta_0 + \beta_1 x_{i,1} + \beta_2 x_{i,2} + \ldots + \beta_n x_{i,n} \tag{5.5}$$

where Deviation degree(i) refers to the deviation degree of Case i.

Deviant cases should be selected from those with high deviation degree. In the context of case selection, deviation degree and typicality are opposite to each other.

5) Influential case selection

An influential case refers to a case that may seem to overturn or question a theory at first sight, but may actually prove the theory instead. There are two ways for determining influential cases, namely hat matrix and Cooke distance (Allen 2007). In hat matrix, the values of the independent variables for all cases are represented by the Matrix **X**, and the values of the dependent variables for all cases are represented by the Matrix **Y**. The expression for the hat matrix **H** is as below:

$$\mathbf{H} = \mathbf{X}(\mathbf{X}^T\mathbf{X})^{-1}\mathbf{X}^T \tag{5.6}$$

$$\mathbf{Y} = \begin{bmatrix} y_1 \\ y_1 \\ \vdots \\ y_n \end{bmatrix} \tag{5.7}$$

$$\mathbf{X} = \begin{bmatrix} x_{1,1} & x_{1,2} & \cdots & x_{1,k} & \gamma_1 \\ x_{2,1} & x_{2,2} & \cdots & x_{2,k} & \gamma_2 \\ \vdots & \vdots & \ddots & \vdots & \vdots \\ x_{n,1} & x_{n,2} & \cdots & x_{n,k} & \gamma_n \end{bmatrix} \tag{5.8}$$

where $x_{n,k}$ refers to the kth independent variable for the nth case; y_n refers to the dependent variable for the nth case; γ_n refers to the constant for the nth case; k refers to the total number of independent variables; and n refers to the total number of cases.

A leverage value is used to determine whether a case is an influential case or not. The leverage value for Case α is determined by the number at position (α, α) in the hat matrix. If the leverage value for a case is greater than $2(k+1)/n$, the case can be considered as an influential case (Gerring 2017).

Cooke distance measures how much the overall results will change if a particular case is removed from the analysis (Seawright and Gerring 2008). The formula for Cook distance is:

$$\text{Cook distance} = \frac{r_\alpha^2 \mathbf{H}_{\alpha,\alpha}}{(k+1)(1-\mathbf{H}_{\alpha,\alpha})} \tag{5.9}$$

where $\mathbf{H}_{\alpha,\alpha}$ refers to the leverage value for Case α; r_α^2 refers to the regression residual for Case α; and k refers to the number of independent variables.

6) Critical case selection

A critical case is one in which a theoretical prediction is most or least likely to be fulfilled. Critical case can be divided into confirmatory critical case and negative critical case. Confirmatory critical case is used to verify the seemingly impossible phenomena predicted by a theory. Negative critical case is used to deny a theory by providing evidence that is contrary to the phenomena.

7) Pathway case selection

A pathway case is one in which a causal effect can be isolated from other potential confounding factors. Pathway cases focus on a single causal factor denoted by X_1, and other factors are uniformly denoted by X_2 (Gerring 2017). For cases with binary variables, there are a total of eight relevant case types, as shown in Figure 5.2.

In Figure 5.2, Case H is the pathway case as the outcome Y that occurs is only related to the factor X_1 of interest. If X_2 is a unidimensional vector, then the total number of combinations is 8 (i.e., 2^3), and if X_2 is an n-dimensional vector, then the total number of combinations is 2^{n+2}.

For cases with continuous variables, we can use the residuals to calculate the path value for a case as below:

$$Y = X_2 + \text{Constant} + \delta_c \tag{5.10}$$

$$Y = X_1 + X_2 + \text{Constant} + \delta_s \tag{5.11}$$

$$PV = \delta_c - \delta_s, \text{ if } \delta_c > \delta_s \tag{5.12}$$

where δ_c refers to the absolute value of residual error for each case in Formula (5.10) and δ_s refers to the absolute value of residual values for each case in Formula (5.11).

	Case A	Case B	Case C	Case D	Case E	Case F	Case G	Case H
X_1	1	0	0	0	1	1	0	1
X_2	1	0	0	1	0	1	1	0
Y	1	0	1	1	0	0	0	1

Figure 5.2   Pathway cases with binary causal factors.

8) Most-similar case selection

A most-similar case selection should contain at least two cases. The most-similar cases are divided into two types, namely hypothesis-generating and hypothesis-testing. In the hypothesis-generating type, researchers need to find cases that are different in results but similar in various factors that may lead to the results (Gerring and Cojocaru 2016). Researchers hope to find one or more key factors that caused this result through in-depth study of these cases, as shown in Figure 5.3(a). In the hypothesis-testing type, researchers need to find cases that behave differently in some factors but the same in others, as shown in Figure 5.3(b).

9) Most-dissimilar case selection

In most-dissimilar cases, only one independent variable (X_1) and the dependent variable (Y) have the same value, and the other independent variables $(X_2, X_3,..., X_k)$ have different values, as shown in Figure 5.4.

Most-dissimilar case is mainly used for exploratory and explanatory case studies, for instance, most-dissimilar case can be used to eliminate necessary causes or to provide weak evidence of the existence of a causal relationship (Seawright and Gerring 2008).

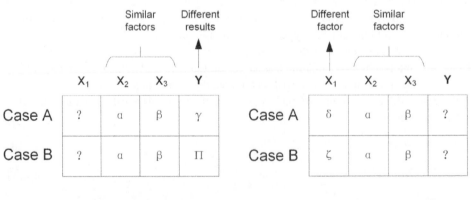

(a) Hypothesis-generating type   (b) Hypothesis-testing type

Figure 5.3   Two case types in the most-similar case selection.

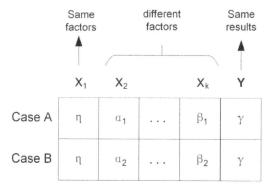

Figure 5.4 The most-dissimilar analysis of the two cases.

5.6.5 Single-case versus Multiple-cases

A single-case study is appropriate primarily for verifying a theory or disproving a theory. It can also be used to investigate an unexplored phenomenon and develop a theory. Multiple-case study is to provide a rich description of the context in which the phenomenon occurs. It is useful for examining phenomena that may occur in a variety of situations. It is also useful for highlighting recurring patterns or finding counterexamples that contradict the defined theoretical constructs. However, in a multiple-case study, it is not easy to determine the number of cases to be studied and how far the researchers should go in collecting data in order to develop a theory. The number of cases to be studied should be limited, while the study still provides sufficient answers to the research questions.

5.6.6 Data Sources

There are five sources of data for case studies: documents, archival records, physical evidence, interviews, and observation (artefacts) (Yin 2017). While the first three already exist in the cases, the last two sources are, in fact, actions to obtain data, i.e., data collection.

1) Documents
   Documented data are relatively stable, natural, and authentic, detailed and easy to quantify, but have the disadvantage of being less retrievable and subject to potential recording errors (Li et al. 2020).
2) Archival records
   Archival records are usually in the form of computer files and records, such as service records, organizational records, maps, and survey information. Most archives are recorded with some purpose and for a specific audience; researchers need to be aware of this to prevent data bias (Li et al. 2020).
3) Physical evidence (artefacts)
   Physical evidence includes physical or cultural artefacts, such as technical devices, tools, artworks. Physical evidence can be collected and observed as part of a divisional visit (Yin 2017).

4) Interviews

Interview is an important source of data for case study. The interview data are targeted, allow for questions to be set according to the objectives of the research, and allow for insight to be gained.

5) Observation

Observation is divided into direct observation and participatory observation. Direct observation is a method in which a researcher simply views and records what is happening or what a person is doing. A researchers in participant observation is not simply a passive observer, but one who takes on different specific roles and is actually involved in the behaviour under study (DeWalt and DeWalt 2010).

5.6.7   Data Collection

After selecting cases and knowing where the data are located or sourced, the next step is naturally to collect data from the cases. The instruments for data collection in case study research are the same as those that have been discussed in the relevant chapters: observations, interviews, document studies, surveys, focused group workshops, and so on. The data collected could be qualitative or quantitative types depending on the research needs, research aims, data sources, and methods of collecting data.

To collect sufficient and credible data in a case study, researchers should not only be good observers but also have good interpersonal skills. The researchers' relationships with the data providers and the informants are crucially important and key to the success of a case study. There are several points to be kept in mind when collecting data (Gagnon 2010; Stake 1995):

- Researchers should establish trust relationships with case study participants.
- Researchers should be observant and practice active listening, and always be alert and avoid presuming what words or deeds will be worthy of attention.
- Evidence should be collected systematically. It should be clear how the data from various sources contribute to the overall aims of the study.
- Multiple sources should be used when collecting data, so that the researchers can analyse a variety of information and strengthen construct validity.
- If changes need to be made to the data collection procedure, these changes must be recorded and documented.
- Researchers should have ways of displaying the progress of the study using, for example, a matrix of tasks to accomplish or keeping track of time spent on different tasks.
- Researchers should have a formal data storage system. Data should be stored in formats that can be referenced so that the patterns of information are clear.
- Researchers should look for causal factors from the information collected.

5.6.8   Data Sorting and Coding

As data are being collected, researchers can organize and sort the data to facilitate analysis. The first thing is to make sure that the collected data are relevant to the study, are in an appropriate format for coding;], and to have the required basic information on their source and how they were collected. Then the data should be organized and classified to make them easier to analyse. The data coding and classification process consists of identifying and coding passages in the texts that describe or relate to categories or concepts connected to the phenomenon of interest.

It is always preferable to carry out the data collection, sorting, cleaning, and analysis iteratively, rather than devote yourself exclusively to data collection for a time and then analyse the data. At that stage, it can be difficult, if not impossible, to fill in gaps or test new hypotheses that suggest themselves in the course of the analysis.

5.6.9   Reliability and Validity

1) Reliability

Reliability is concerned with the question of whether the results of a study are repeatable. Reliability can be divided into two types: internal reliability and external reliability. *Internal reliability* means that other researchers would arrive at essentially the same findings if they were to analyse and interpret the data produced by the study. In other words, the conclusion drawn from the evidence by multiple independent observers and coders would be sufficiently consistent to describe the phenomenon in a similar way. To enhance internal reliability, it is recommended that the researchers take the following five steps (Gagnon 2010):

1) Use concrete and precise descriptors of the entire process.
2) Safeguard raw data.
3) Involve several researchers if time and budget permits.
4) Confirm the collected data with the source.
5) Have the interpretation of the data reviewed by peers.

External reliability means that an independent researcher using the same methodology would obtain essentially the same data and discover the same phenomenon if he or she were to observe the same environment or a similar environment. External reliability can be significantly enhanced if five major threats are addressed through the steps described below (Gagnon 2010).

1) Precisely establish researcher's position describing to what extent they are part of the phenomenon they are studying.
2) Describe the informant selection process.
3) Describe the environment of data collection process.
4) Clearly define the study's concepts, constructs, and units of analysis.
5) Describe data collection strategies.

2) Validity

Validity addresses the integrity of conclusion generated from the case study. Validity is established by producing a valid and rigorous interpretation of the evidence. Validity consists of the following three aspects: internal validity, external validity, and construct validity. Internal validity makes sure that the description of the phenomenon is an accurate representation of the observed reality. For example, if we suggest that X causes Y, can we be sure that it is X that is responsible for variation in Y and not something else that is producing an apparent causal relationship (Bryman 2016)? External validity is concerned with whether the result of a study can be generalized beyond the specific research context (Bryman 2016). To maintain external validity, researchers have to consider the factors that threaten the comparability and transferability of the results. Construct validity means selecting the most appropriate measurement tool for the concepts being studied. It refers to the question of whether a measure that is devised of a concept does reflect the concept that it is supposed to be denoting (Bryman 2016). The steps for ensuring internal validity, external validity, and construct validity are presented in Figure 5.5.

5.6.10 Data Analysis and Interpretation

After data collection is completed, researchers need to analyse and search for patterns that may emerge from the data. To do so, researchers should get immersed in the evidence, in the configuration of the facts and the interconnections. A configuration is a specific combination of factors (or stimuli, causal variables, determinants, etc.) that produces a given outcome of interest (Rihoux and Ragin 2009). Usually, the important meaning will come from reappearance over and over (Stake 1995). Sometimes, significant meaning can be found in a single instance.

Data interpretation is the most important and challenging part of a case study, as it requires the researchers' analytical power and creativity. The researchers should start with a detailed description of each case, reflect on its implications, analyse the collected data and related patterns from a wider perspective, and form his or her own initial overview. This process prompts a search for new concepts and theories in the evidence, to achieve a better understanding of the perception, behaviour, or situation, and to arrive at a description or explanation of the phenomenon of interest (Flyvbjerg 2011).

The development of explanatory schemes in the data interpretation stage yields explanations through a back-and-forth movement between generating ideas and checking them against the data (Bazeley 2013). The aim here is to determine the extent to which the data from a case can support the hypothesis proposed by the researcher. A hypothesis that is not

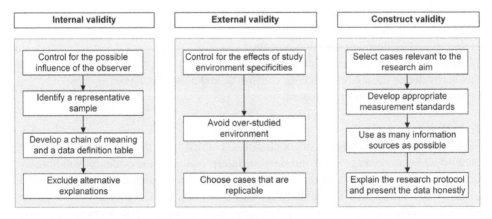

Figure 5.5 Steps for ensuring internal validity, external validity, and construct validity.

supported by the evidence is not valid, and a hypothesis that is not adequately supported should be revised (Yin 2012). When a proposed explanation has passed the test of real case data, it should be compared with what is in the literature. The aim here is to contribute to the theory by identifying and analysing any discrepancy between the proposed explanation and the existing theory (Gerring 2017). If there is some support in the literature, this supports the internal validity of the study and makes it possible to generalize the results (Thomas 2021).

The ways of linking collected data to the research hypothesis include pattern matching, logic model, constructive interpretation, time-series analysis, and cross-case clustering analysis (Yin 2017). These analytical techniques help researchers assess and enhance the quality of case study.

Pattern matching is the most common technique used in case study. Pattern matching matches patterns based on empirical evidence (patterns based on research findings) with patterns based on predictions (one or more possible predictions) (Trochim 1989). By using pattern matching, researchers can demonstrate whether the findings reflect the hypothesis.

Logic model is used to reconstruct complex phenomena to reflect all aspects of the case. A logic model is a complex and precise chain of phenomena over a certain period of time. These phenomena can show a repetition and cycle of 'cause – effect – cause – effect', where the dependent variable of the previous stage becomes the independent variable of the next stage.

Time series in most cases contain only one independent or dependent variable. The logic inherent in the design of a time series is to compare the information with three trends: (i) a trend assumed based on theoretical analysis prior to the beginning of the case study; (ii) an opposite trend identified in the early phase of the case study; and (iii) any trend that undermines the internal validity of the case study (Yin 2017).

Cross-case cluster analysis, as the name implies, examines more than two cases. Documentation tables can be used to present individual case information. A qualitative analysis of the documentation tables allows for a conclusion to be drawn across different cases (Yin 2017). Then, a framework can be developed to present the analytical results and conclusion. There are four criteria for interpreting the finding of a case study (Yin 2017): (i) considering all information as far as possible; (ii) considering all reasonable competing explanations; (iii) stating the most meaningful aspects of the case; (iv) having the sound use of the researcher's expertise. There are four strategies of cross-case cluster analysis: (i) developing a case description; (ii) working the data from the ground up; (iii) relying on theoretical hypothesis; (iv) examining plausible rival explanations. Through simultaneous data collection, analysis, and interpretation processes, the researchers may modify or replace initial research problem if needed.

5.7 Case Study Report

Case study report is the final product that presents the findings of a case study. There are six structures that can be used in case study report, namely linear-analytic structure, comparative structure, chronological structure, theory-building structure, suspense structure, and unsequenced structure (Yin 2017).

1) **Linear-analytic structure**

   Linear-analytic structure is a standard structure for composing case study reports. Similar to a journal article, it starts with research problem and literature review, then presents the methods, findings, implication of the findings, and conclusions. Linear-analytic structure is suitable for all types of case studies.

2) **Comparative structure**

   A comparative structure repeats the same case two or more times, comparing different descriptions and interpretations of the same case. The same case can be described and explained repeatedly from different models to show the degree to which the evidence fit each model. Comparative structure is suitable for all types of case studies.

3) **Chronological structure**

   A chronological structure presents the case study evidence in chronological order. For example, presenting the case history in the order of its early, middle, and late phases. Chronological structure is suitable for all types of case studies.

4) **Theory-building structure**

   A theory-building structure follows theory-building logic to present a case study. This structure is suitable for exploratory and explanatory case studies.

5) **Suspense structure**

   In a suspense structure, the finding of a case study and its significance are presented at the beginning. The remainder is devoted to the explanation of this finding. Suspense structure is suitable for explanatory case study.

6) **Unsequenced structure**

   For unsequenced structures, the order in which the various parts of the case study report are presented is not particularly important. Even if the order is changed, the value of the description is not affected. However, researchers need to be aware of integrity when using this structure. Unsequenced structure is suitable for descriptive case study.

5.8 Qualitative Comparative Analysis (QCA) Method

In case study, the subject of study often involves multiple cases and the individual cases may not completely meet the research needs. At the same time, in a complex social phenomenon, there may be multiple and concurrent combinations of causal variables that make it difficult for quantitative statistical analysis to provide valid conclusions. Qualitative comparative analysis (QCA) has been widely used because of its ability to effectively and systematically process data for multiple-case comparisons.

QCA is a case study-oriented research method to examine the relationships of necessary and sufficient conditions to the outcomes (Ragin 2014). In other words, it is a Boolean algebra-based approach to set-theoretic group state analysis that examines the sufficient and necessary subset of relationships between antecedent conditions and outcomes, enabling a holistic exploration of how complex social problems induced by multiple concurrent causes and effects (Rihoux and Ragin 2009). QCA goes beyond the traditional case study approach to systematically examine the causes of a phenomenon and the interactions and combinations of possible relationships between internal generating factors, trying to explain the key factors that contribute to the phenomena, the interconnections between factors and the complex combinations of causes that inspire the phenomena, with the aim to deepen the understanding of the complex causal relationships that generate phenomena. QCA's typical strengths are its ability to handle large, medium, and small samples and to analyse complex configuration (i.e., any multidimensional cluster of co-occurring, conceptually distinguishable features) problems (Meyer et al. 1993; Miller 2018). The overall process of conducting QCA is:

1) **Condition identification and model construction.** QCA emphasizes the identification of conditions relevant to the research questions based on theoretical or empirical knowledge. QCA researchers need to argue for the association of each condition with the outcome from a configuration perspective. A key factor to consider in constructing a configuration model is the number of conditions, which is determined by considering both the sample size and the parsimony of the model. For example, as there are theoretically *2n* combinations of *n* conditions, an excessive number of conditions could easily lead to the number of configurations exceeding the number of observed cases, thus creating the 'problem of limited diversity' of cases. Furthermore, although large sample QCA may avoid the problem of limited diversity, an excessive number of conditions may lead to complication in the interpretation of the results.

2) **Case selection.** Case selection and condition identification often occur simultaneously or iteratively, with no fixed order of precedence. Researchers need to select relevant cases based on the research questions. QCA was originally developed for small and medium sample size research contexts, which require researchers to develop a closer and more intimate relationship with the case data. Studies using small sample sizes should follow the principles of theoretical sampling, selecting samples based on the characteristics of the theory and the cases to ensure sufficient heterogeneity between cases for comparison (Ragin 2014). There are also studies that show the potential of QCA in large sample studies. For example, Greckhamer et al. (2008) showed that QCA can handle even thousands of cases. Studies using large sample sizes can employ more traditional random sampling strategies or purposive sampling methods (Fiss 2011).

3) **Calibration of conditions and outcome variables.** Calibration is the process of assigning a specific conditional set affiliation to a case. Only after the original case data have been calibrated to a set affiliation score can further subset relationship analysis of necessity and sufficiency be performed. Depending on the set form, QCA applications are divided into two main categories: clear sets (csQCA) and fuzzy sets (fsQCA). Of these, csQCA is a dichotomous transformation of variables to '0' or '1'. The fsQCA, on the other hand, allows for partial affiliation scores between '0' or '1', i.e., assessing the degree of affiliation of a condition between 'fully affiliated' and 'not affiliated at all'. The degree of subordination between 'fully subordinate' and 'not subordinate' (Li and He 2015).

For fuzzy set calibration, there are two main calibration methods used in the existing studies. One is the indirect calibration method, where researchers use their judgement to assign multiple values between '0' and '1' to each condition. The second is the direct calibration method, where researchers propose three qualitative anchor points based on theory or practice: complete affiliation, complete disaffiliation, and intersection, which are then calibrated using the algorithms provided by a software. Of these, the direct calibration method uses a statistical model, which is more formalized and is the most commonly used calibration method. It is important to note that the choice of anchor points should follow the principles of rationality and transparency, and that researchers can refer to the existing theory or provide theoretical justification, or refer to external samples for empirical evidence, or choose anchor points based on the frequency distribution of the sample data.

4) **Necessary conditions analysis.** Before carrying out the standard analysis for QCA, researchers should check whether any condition is necessary for the result. A 'necessary condition' means that the condition is always present when the result is present, in other words, the result cannot be generated without it. The researchers can perform a necessity analysis using a procedure provided by a software to assess the relationship between the set of results and a subset of the set of conditions. It is usually determined that the necessity conditions need to achieve a consistency score of 0.9 and have sufficient coverage (Ragin 2014). It is also important to distinguish between the necessity of a condition and the generality of a condition; when a condition appears in each of the final configurations, it may appear to be necessary, but the existence of its necessity cannot be inferred from these findings.

5) **Configurational analysis and interpretation of results**. Configurational analysis consists of two steps: truth table refinement and standard analysis. First, the researcher sets relevant thresholds to initially screen the truth table rows according to three criteria. The first criterion is to determine the minimum number of cases to avoid empirically trivial configurations. In QCA with small number of cases (e.g. 10–40 cases), the number of cases in each type of configuration analysis may be set as one or two. In QCA with larger number of cases (e.g. >40 cases), the number of cases in each type of configuration analysis can be higher. The second criterion is the determination of a consistency threshold to ensure the strength of interpretation of the configuration, with existing studies indicating consistency greater than 0.8 as the minimum acceptable standard (Ragin 2014). The third criterion relates to proportional reduction in inconsistency (PRI) consistency, and the best practice suggests keeping the PRI above 0.75 to avoid the problem of 'simultaneous subset relationships' (Greckhamer et al. 2018).

After retaining the rows that satisfy the three analysis criteria for standard analysis, three types of solutions are output: complex, parsimonious, and intermediate. Complex solutions are based on the original data, without any counterfactual analysis, and usually contain more groupings and antecedents. Parsimonious solutions are based on simple but difficult counterfactual analysis and contain a minimum number of states and conditions. Intermediate solutions consider only simple counterfactuals, incorporating logical residuals that are consistent with theoretical expectations and empirical evidence. Reasonable and moderately complex intermediate solutions are often the preferred choice for reporting and interpretation in QCA.

The presentation of QCA results is generally done using the approach recommended by Fiss (2011). A common approach to interpreting the results of analysis is to use intermediate solutions to determine the number of configurations leading to the results and the conditions under which these configurations are included, and then use the results of the parsimonious solution to identify the core conditions that are more important for a given configuration (Fiss 2011). Conditions that appear in the parsimonious solution are referred to as core conditions for a given grouping, indicating a strong causal relationship with the outcome of interest. The remaining conditions that appear in the intermediate solution but not in the parsimonious solution are called marginal conditions and have a weaker causal relationship with the outcomes.

6) **Robustness test.** The robustness test applied to QCA focuses on parameter sensitivity, measurement error, and model specification (Baumgartner and Thiem 2017; Rohlfing 2018). Robustness test for QCA should examine whether the configurations vary significantly under different parameters (e.g., PRI and the number of cases in each type of configuration analysis). The configurations under different parameters can be considered robust if it has a similar combination of conditions, consistency, and coverage as the results obtained prior to the robustness test (Schneider and Wagemann 2012).

5.9   Relationships between Qualitative Research, Quantitative Research and Case Study

The relationships between a case study method with quantitative research and with qualitative research can be seen in Figure 5.6, where the dotted circle represents case study methods. Whether conducting a single case study or a multiple-case study, researchers should not limit themselves to qualitative research methods, especially with multiple-case study with large samples, where it is impossible to deal with them only in a qualitative way since there are so many cases. Gerring (2017) pointed out that purely narrative case studies without any numerical analysis or purely quantitative case studies without any textual description are basically nonexistent.

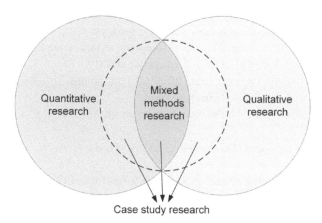

Figure 5.6   Relationships between qualitative research, quantitative research, case study, and mixed methods research.

5.10   Summary

This chapter is perhaps one of the most complicated chapters presented in this book. It started with a discussion on the basic definitions and philosophical paradigms of case study. It then discussed the types of research questions or phenomenon that are suitable to use case study as a research method. There are 10 steps in case study implementation: research problem identification, research hypothesis development, purpose of case selection, methods for selecting case, single-case versus multiple-cases decision making, data source, data collection, data sorting, data coding, reliability and validity testing, and data analysis and interpretation. Case study report is also discussed in detail, as there is specific framework, structure, and contents to be followed and included. Next, we discussed QCA, as it is often used in case study research. Finally, we discussed the relationships between qualitative, quantitative, mixed methods, and case study research and designed a diagram (Figure 5.6) to show the relationships.

Review Questions and Exercises

1   What situation is suitable for using case study methods?
2   How do you decide whether to use single case or multiple-case study?
3   What are the different types of case studies and what are the strengths of each type?
4   What are the processes of undertaking case study research?
5   How do you select cases?
6   What are the methods can be used for case study analysis?
7   How do you draw conclusions in case study research?

References

Allen, M.P. (2007). *Understanding Regression Analysis*. New York, NY: Springer.

Baumgartner, M. and Thiem, A. (2017). Often trusted but never (properly) tested: evaluating qualitative comparative analysis. *Sociological Methods & Research* 46: 345–357.

Bazeley, P. (2013). *Qualitative Data Analysis: Practical Strategies*. SAGE Publications.

Bryman, A. (2016). *Social Research Methods*. Oxford University Press.

Crowe, S., Cresswell, K., Robertson, A. et al. (2011). The case study approach. *BMC Medical Research Methodology* 11 (1): 1–9.

DeWalt, K.M. and DeWalt, B. (2010). *Participant Observation: A Guide for Fieldworkers*, 2e. Lanham, MD: Alamira Press.

Fiss, P.C. (2011). Building better causal theories: a fuzzy set approach to typologies in organization research. *The Academy of Management Journal* 54 (2): 393–420.

Flyvbjerg, B. (2011). Case study. *The Sage Handbook of Qualitative Research* 4: 301–316.

Gagnon, Y.C. (2010). *The Case Study as Research Method: A Practical Handbook*. Presses de l'Universite du Quebec (PUQ).

Gerring, J. (2017). Qualitative methods. *Annual Review of Political Science* 20: 15–36.

Gerring, J. and Cojocaru, L. (2016). Selecting cases for intensive analysis: a diversity of goals and methods. *Sociological Methods & Research* 45 (3): 392–423.

Greckhamer, T., Furnari, S., Fiss, P.C., and Aguilera, R.V. (2018). Studying configurations with qualitative comparative analysis: best practices in strategy and organization research. *Strategic Organization* 16 (4): 482–495.

Greckhamer, T., Misangyi, V.F., Elms, H., and Lacey, R. (2008). Using qualitative comparative analysis in strategic management research: an examination of combinations of industry, corporate, and business-unit effects. *Organizational Research Methods* 11 (4): 695–726.

Li, L., Liu, Y., and Feng, Y.C. (2020). *Case Study Research: Methods and Practice*. Peking University Press.

Li, W. and He, H.B. (2015). The logic and application of qualitative comparative analysis in social science research. *The Journal of Shanghai Administration Institute* 5: 92–100. (in Chinese).

Meyer, H.D., Manthe, U., and Cederbaum, L.S. (1993). The multi-configuration Hartree approach. In: *Numerical Grid Methods and their Application to Schrödinger's Equation* (eds. C. Cerjan), 141–152. Dordrecht: Springer.

Miller, D. (2018). Challenging trends in configuration research: where are the configurations? *Strategic Organization* 16 (4): 453–469.

Ragin, C.C. (2014). *The Comparative Method: Moving beyond Qualitative and Quantitative Strategies: With a New Introduction*. Oakland, California: ISBN 978-0-520-95735-0. OCLC 881322765.

Rihoux, B. and Ragin, C.C. (2009). *Configurational Comparative Methods: Qualitative Comparative Analysis (QCA) and Related Techniques*. SAGE Publications.

Rohlfing, I. (2018). Power and false negatives in qualitative comparative analysis: foundation, structure and estimation for empirical studies. *Political Analysis* 26: 72–89.

Schneider, C.Q. and Wagemann, C. (2012). *Set-Theoretic Methods for the Social Sciences*. Cambridge: Cambridge University Press.

Seawright, J. and Gerring, J. (2008). Case selection techniques in case study research. *Political Research Quarterly* 61 (2): 294–308.

Stake, R.E. (1995). *The Art of Case Study Research*. SAGE Publications.

Thomas, G. (2021). *How to Do Your Case Study*. SAGE Publications.

Trochim, W.M.K. (1989). Outcome pattern matching and program theory. *Evaluation and Program Planning* 12 (4): 355–366.

Yin, R.K. (2012). Case study methods. In: *APA Handbook of Research Methods in Psychology, Vol. 2. Research Designs: Quantitative, Qualitative, Neuropsychological, and Biological* (eds. H. Cooper, P.M. Camic, D.L. Long, et al.), 141–155. American Psychological Association.

Yin, R.K. (2014). Getting started: how to know whether and when to use the case study as a research method. *Case Study Research: Design and Methods* 5: 2–25.

Yin, R.K. (2017). *Case Study Research: Design and Methods*. Thousand Oaks, CA: SAGE Publications.

6

Technology-Enabled Experimental Research

6.1 Introduction

Experimental research is crucial to both natural and social science research. There are two types of experiments, natural experiment and social experiment. Natural experiment uses instruments and equipment to change, control, or simulate the object of research, while eliminating external influences, to make something happen, in order to understand a natural phenomenon, nature, and natural laws. Social experiment aims to see how people behave in different situations or how they respond to different policies. In recent years, technology-enabled experimental research methods have become popular in social and management research. In particular, the convergence of neuroscience and technology is facilitating development in natural science and social science that not so long ago would have seemed unthinkable, as is eye-tracking technology. This chapter first provides an overview of experimental research, then discusses neurotechnology and eye-tracking technology as two emerging technologies, going on to develop a methodological framework for using these technologies to enhance experimental methods in human-related research. Relevant issues such as research scenarios, research variables, research ethics, costs, and skills are also discussed.

6.2 Overview of Experimental Research

Generally, the term 'experimental research' refers to the methods developed for the specific purpose of testing causal relationships (Dane 1990). It is a scientific method of conducting research using independent and dependent variables and observing an individual or a group experiencing a certain condition, or individuals or groups under different conditions. In experimental research, one or more independent variables are manipulated and applied to one or more dependent variables to measure the effect of independent variables on dependent variables.

Research Methodology and Strategy: Theory and Practice, First Edition. Patrick X.W. Zou and Xiaoxiao Xu.
© 2023 John Wiley & Sons Ltd. Published 2023 by John Wiley & Sons Ltd.

6.2.1 Conditions of Experimental Research

Before starting experimental research, researchers need to clarify the following concepts (Dane 1990):

1) Temporal priority: experimental research is based on the requirement that the suspected cause precedes the effect. In other words, something that will happen tomorrow cannot cause something that happens today. The most common way to establish temporal priority is to manipulate the independent variable that is the suspected cause under consideration.
2) Control over variables: experimental research involves exerting some control over the research environment, so that the independent variable and dependent variable may be accurately measured.
3) Random assignment: random assignment is the procedure that provides all participants an equal opportunity to experience any given level of the independent variable. The aim is to eliminate the impact of other possible causes as much as possible.
4) Alternative explanation: alternative explanations should be eliminated to make valid inference from experimental research.

6.2.2 Types of Experimental Research

There are three main types of experimental research: pre-experimental research, true experimental research, and quasi-experimental research.

Pre-experimental research is an observational approach to performing an experiment. It allows little or no control over extraneous variables that might be responsible for outcomes other than what the research intended (Salkind 2011). There is no control group and the participants are selected randomly from a population. Pre-experimental research can be divided into three categories.

1) One-shot case study design: subjects are assigned to one group and are tested on some dependent variables after an intervention has taken place.
2) One-group pretest posttest design: researchers apply a test on some dependent variables both before and after an intervention. This provides a comparison of performance with or without intervention to investigate the effects of the invention on the subjects.
3) Static group comparison design: researchers apply a test at the end of the process to compare the results from the subjects.

True experimental research controls selection of subjects, assignment to groups, and assignment to treatment. It includes steps in selecting and assigning subjects in a random fashion, plus a control group, thereby lending a stronger argument for a cause-and-effect relationship (Salkind 2011). Random fashion involves random selection of participants, random assignment of treatments, and random assignment of groups. A typical 'true' experimental research design is 'pretest posttest control group design' with the following steps:

1) Randomly assign the subjects to the experimental group or the control group.
2) Pretest each group on the dependent variable.

3) Apply the treatment to the experimental group while the control group does not receive the treatment.
4) Posttest both the experimental group and the control group on the dependent variable (in another form or format, if necessary).

In quasi-experimental research, the hypothesized cause of difference between groups has already occurred, which means preassignment to groups has already taken pace. Quasi-experimental research allows for the exploration of questions that otherwise could not be ethically investigated (Salkind 2011) for reasons such as health and safety risks.

6.2.3 Experimental Research Design

Experimental research design involves the determination of the number and arrangement of independent variables and dependent variables and their levels in a research project. There are four types of experimental design (Dane 1990).

1) Basic design. This can be considered a posttest-only control group design. Subjects are assigned to two groups at random and only one group has an intervention. Researchers perform tests at the end of the process. Coupled with careful control over procedures, this type of design effectively provides a comparison between control and intervention conditions by comparing the dependent variable responses of the two groups.
2) Solomon four-group design. Subjects are assigned to one of four randomly allocated groups. These groups provide all four possible permutations for posttest or pre- and posttest control groups and control and noncontrol groups.
3) Factorial design. This includes more than one independent variable, which could test the interactions between independent variables. An interaction occurs when the effect of one variable depends on which level of another variable is present.
4) Repeated measures design. A = specific factorial design in which the same participants are exposed to more than one level of an independent variable.

6.3 Electroencephalography Technology

Cutting-edge technologies are continually emerging, such as the Internet of Things, electroencephalography (EEG) technology, eye-tracking technology, virtual reality, augmented reality, and sensing technology. These are changing the precision of experiments and simplifying the process. The following sections explore experimental research enabled by EEG and eye-tracking technologies.

6.3.1 Description of Electroencephalography Technology

Neurotechnology belongs to a discipline within biomedical engineering that uses engineering techniques to explore neural systems (Hetling 2008). The most common neurotechnology used in neural engineering is electroencephalography. EEG is a noninvasive measurement that records and analyses electric signals generated by the neurons to detect electrical activity inside the human cortex (Biasiucci et al. 2019). Previous studies have found that EEG signals are very closely linked to human mental workload (Pfurtscheller and Da Silva 1999). There are five main frequency bands of signals: delta waves (1–3 Hz), theta waves (4–7 Hz), alpha waves (8–12 Hz), beta waves (13–30 Hz), and gamma waves (31–50 Hz) (Wang et al. 2017). Delta waves generally occur during sleep and are related to a state of unconsciousness. Theta waves usually appear during relaxed wakefulness (Levin and Chauvel 2019). Alpha waves can be seen in the EEG during a normal wakeful state, while beta waves are shown when a person is alert and thinking actively (Kirstein 2007). Gamma waves are the fastest brain waves; they can be found during information processing and mentally demanding activities (Buzsaki 2006). A schematic diagram of EEG signal data collection is shown in Figure 6.1. Through EEG-enabled experimental tests, the subjective bias of traditional survey-based assessments of mental and emotional states can be overcome (Hwang et al. 2018).

6.3.2 Application of EEG Technology

Neurotechnology has been applied to several research fields, such as architecture (Erkan 2018), tourism (Brouder 2014), and construction management (He et al. 2019). EEG equipment consists of a number of electrodes for acquiring brain waves from different areas of the brain, each corresponding to the activity of that part and therefore to the function of the brain (Teplan 2002). Researchers usually use EEG equipment to investigate emotional states, occupational stress, workers' attention and vigilance, as discussed in the following sections.

Workers' emotional states are critical factors affecting work performance (Hwang et al. 2018). Since individuals' emotions are associated with physiological activities, EEG was applied to measure human emotions by assessing human physiological responses (Chanel et al. 2011). Hwang et al. (2018) used a wearable EEG sensor to measure workers' valence and arousal levels during their ongoing tasks. Guo et al. (2017) applied a wearable technology-based method to examine the correlations between workers' psychological status and

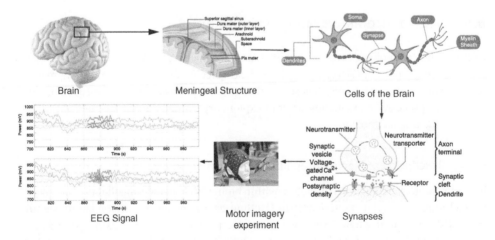

Figure 6.1 Schematic diagram of EEG signal data collection.

physical status. Xing et al. (2019) proposed a multicomponent and neurophysiological intervention for the emotional and mental states of high-altitude workers.

Occupational stress is defined as the negative physical and mental reactions that arise when the demands of the job exceed the worker's capabilities, which can be measured by EEG (Jebelli et al. 2018). Kim et al. (2004) found the ability of the EEG to distinguish between different levels of arousal, defined as the state of being aroused and the degree of attention, and to correlate closely with stress through changes in response to different stressors.

When workers are engaged in repetitive tasks, they may lose sufficient vigilance and attention to the hazards around them (Wang et al. 2019). Some researchers used brain waves to measure workers' attention and vigilance. Wang et al. (2017) proposed a wireless and wearable EEG system to quantitatively and automatically assess the workers' attention levels by processing signals from the human brain. Ke et al. (2021) investigated the correlation between brain activity and distraction via an EEG device, and the results showed that beta and gamma waves in the left temporal and right prefrontal cortex could distinguish focused and distracted statuses.

6.4 Eye-tracking Technology

6.4.1 Description of Eye-tracking Technology

Eye-tracking technology is an effective tool for monitoring and analysing a subject's visual search (Jeelani et al. 2018). The human eye is limited in its ability to notice everything around it at once (Nilsson 1989), and the subject's attention is directed to where their eyes are focused (Yarbus 1967). Eye-tracking technology has been applied in healthcare (Henneman et al. 2017), multimedia (Hyönä 2010), safety management (Comu et al. 2021), and transport (Wu et al. 2021). An eye-tracking device can record and track eye positions and eye movement, and output different indicators, such as first fixation time, total fixation time, percentage of dwell time, as well as number of runs recorded from eye movements (Rosch and Vogel-Walcutt 2013).

6.4.2 Application of Eye-tracking Technology

Eye-tracking is good at collecting data in a complex environment through target search and recognition. This technology has numerous potential applications, such as research into attention and mental fatigue. Eye-tracking is widely considered to be the most direct and continuous measure of attention; where a person looks is highly correlated with their focus (Armstrong and Olatunji 2012). For example, Hasanzadeh et al. (2017) investigated the impact of safety knowledge on workers' attentional allocation. Meanwhile, eye-tracking can help assess mental fatigue through instrument-based measurement, as opposed to subjective assessment (Li et al. 2020). For example, Li et al. (2020) applied wearable eye-tracking technology to identify and classify the mental fatigue of equipment operators. Li et al. (2019) assessed the detrimental effects of mental fatigue on the hazard detection performance of excavator operators using eye-tracking techniques.

6.5   Framework for Conducting Technology-enabled Experiment

Figure 6.2 provides a methodological framework for conducting technology-enabled experimental research. The process starts with a defined research problem and aim. The influence mechanism among psychological factors, physical factors, and external factors can be explained using the EEG and eye-tracking tools, which provide objective neurophysiological data to reflect psychological states. Depending on the types of problems to be studied, the neurophysiological measurement metrics are then selected. Following the neurophysiological measurement metrics, the corresponding devices are selected. Meanwhile, the experimental tasks and stimuli are selected based on psychological factors, physical factors, or behaviour to be studied. These selections support the experiment to have real-life scenes or laboratory scenes with wearable and nonwearable devices, and the data collected include neurophysiological data and behavioural data.

The neurophysiological data collected is subject to preprocessing, feature extraction and selection, followed by the analysis of the selected data. Data analysis could either use traditional methods, such as statistical analysis and expert experience, or more recent technology-enabled methods, such as machine learning and big data analytics. The results of data analysis will in turn examine the research problem and aims.

6.6   Research Objectives, Variables, and Complementary with Traditional Methods

EEG and eye-tracking technologies are useful research tools to investigate relationships between variables. This could be between physical factors and psychological factors, or between external factors and psychological factors, or interventional factors and psychological factors, or between psychological factors and behavioural factors, as shown in Figure 6.3. Examples of these factors are listed in Table 6.1.

When applying neurotechnologies, research objectives fall into two main categories: (i) to examine whether the neurophysiological indicators can effectively measure the mental state and (ii) to study the relationships between research variables by measuring mental states with neurophysiological tools. Here, subjective surveys and/or physical indicators representing psychological variables help determine whether EEG and eye-tracking metrics have been effective to measure any psychological variables.

Neurotechnology-based experimental research methods could help collect objective data, leading to more accurate results than 'traditional' methods. This is because it is possible to access information from the subconscious mind that may be difficult to gather by traditional methods. However, it does not mean that conventional surveys and interview methods are to be thrown away. Instead, the two methods may be used together to supplement one another. There may be situations where it is difficult to set up the neurotechnology-based experimental scenarios; in some cases, it would be useful to use two sets of data collected by neurotechnology and traditional survey or interview for cross-comparison; or, the two methods may play different roles and functions at different parts or phases of research. For example, traditional methods such as questionnaire surveys and interviews could be carried out in the pre-investigation phase. When multiple methods are used to collect data, it is necessary to consider how different types of data can be brought together for different analysis.

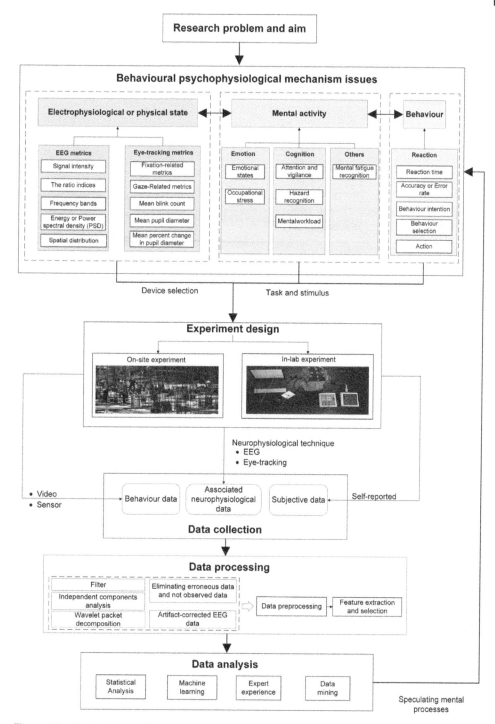

Figure 6.2 Methodological framework of technology-enabled research.

Figure 6.3 Relationships between physical factors, external factors, psychological factors, and behavioural factors.

Table 6.1 Examples of different factors.

Factor category	Factor examples
Physical factors	Physical fatigue, blood pressure, heart rate, breath, and body surface temperature
Psychological factors	Mental fatigue, cognition, emotion, stress, vigilance, and attention
Behavioural factors	Driving behaviour, sleep behaviour, and chaotic behaviour; Body movements (locomotion, speech, and nonverbal communication, flexion, extension, abduction, rotation, and circumduction)
External factors	Noise, temperature, humidity, light intensity, and ventilation
Interventional factors	Fake news, information originality and familiarity, information disclosure, meditation intervention, family nurture intervention, care (body contact, encouragement), mindfulness intervention, and cognitive and affective creativity interventions

6.7 Technology-enabled Experimental Research Design: Scenarios, Variables, and Tasks

A key step in experimental set up is to design the scenarios, tasks, and research variables. These are often simplistic tasks that reflect a game or dilemma, rather than a real-world case or problem, as variables are more difficult to control in a field experiment. This means that it is hard to understand people's psychological states in real-world situations. The advancement of wearable EEG and eye-tracking sensors, which are portable, wireless, and affordable, opens a new door towards the nonintrusive collection of employees' electro-physiological or physical data.

In within-subject design, each subject is assigned to each level when there are two or more independent variable levels. That is, each subject accepts each independent variable level. Since each subject has performed treatment of different conditions, the participants are compared with themselves under changes in conditions, and thus the results are not considered to be caused by individual differences. As such, the results of within-subject design are effective. However, due to the fact that a subject receives different condition treatments before and after, there may be an impact of the former condition on the latter condition. A possible solution is to randomly present experimental conditions to avoid such impact.

The between-subject design means that some subjects are assigned to a level when there are two or more independent variable levels; each subject only accepts one independent variable condition in the between-subjects design, thus the treatment of one condition will not affect the other. The disadvantage here is that individual differences between the groups may affect the results, and therefore between-subject design needs to deal with between-subjects' individual differences.

A usual element of research design, including research involving eye-tracking technology, is the selection of design methods, in this case block design and event design. *Block design* usually includes commutative tasks and control conditions with a fixed time period (block) for each phase. The task or stimulus in an event-related design is in random order and with different time periods. *Event-related design* is mainly used to capture electrophysiological or physical signal data in response to specific events. EEG and eye-tracking-based experiments mostly adopt event-related design, as this design is suited for actual tasks that induce corresponding psychological variables, measurable through EEG or eye-tracking technologies.

6.8 Human Acceptance and Participation

Neurotechnology-based experimental research methods require human acceptance and participation. Given that most participants in the experiment might not be familiar with the equipment and the process, it is practically and ethically necessary to explain processes in full detail so that they feel comfortable. Incentives such as vouchers and lucky draws may encourage more participation.

Moreover, the experimental environment and the real onsite environment could be different. Some neurotechnology experiments can only be carried out in the laboratory and many real-world scenarios are difficult to simulate. In some experiments, the experimental environment is like a surgery operation workstation, which is bound to cause certain psychological pressure to the participants. There are differences between the psychological activities of the participants in the experiment and those in real life, which may lead to errors in the experimental results.

Due to limitations in equipment, it is difficult to recruit a large number of participants for one research. Instead, a limited number may be selected to represent the population, and this must be coupled with the participants' level of cooperation.

6.9 Costs of Technology and Technical Skill

Currently, the acquisition costs for technology (EEG and eye-tracking) equipment remain expensive. For example, the cost of an fMRI device is tens of thousands of dollars. The ongoing costs for equipment operation and maintenance are also high. In addition, cost for participant recruitment and participation could add up to tens of thousands of dollars.

The complexity of technology equipment operation is a challenge for researchers and the training requires a lot of time and financial support. In addition, the complex experiment design requires comprehensive knowledge, and neurotechnology relies on repetitive stimulation to obtain data, which might limit the effectiveness of complex studies. The cost of technology and equipment will doubtlessly decrease as time goes by, and operational skill will be improved as more and more researchers use the technology and equipment.

6.10 Research Ethics

While neurotechnology may have boundless potential for different fields of research, there are ethical risks. Technology itself is neutral and an activity itself depends on its motivation and method. Neurotechnology may be controversial, as it may constitute manipulation or an invasion of privacy. Some methods of neurotechnology are deeply unconscious, that is, tracking an underlying mechanism of cognition that people cannot grasp. It is thus even more necessary to follow ethical principles of transparency, justice and fairness, nonmaleficence, responsibility, privacy, beneficence, freedom and autonomy, trust, sustainability, dignity, and solidarity (Jobin et al. 2019) when developing and implementing neurotechnology-based experimental methods. Self-assessment, organization ethical committee approval, and external assessment and approval processes will all help to ensure research ethics, as well as the full disclosure to participants discussed above.

6.11 Summary

Experimental research is one of the main methods used in natural and social research. However, the subjective nature of research data collected with traditional methods remains as a major concern. In this chapter, we have discussed emerging neurotechnology and eye-tracking technology as new methodologies to overcome subjective data collection. The methodological framework proposed for experimentally applying neurotechnology includes problem definition and description (such as human emotion, cognition, mental fatigue states, and behaviour), technology selection, signal processing, scenario and experimental design, the data collection process, and data analysis process. We have also discussed the challenges in implementing the new experimental methods, such as cost, skill, participation, ethics, and complementary function with traditional methods.

By using the technologies and following the methodological framework proposed, a wide range of psychological factors can be studied, as well as the influence mechanisms among psychological factors, physical factors, and behavioural factors. The use of EEG and eye-tracking technologies helps characterize patterns of electrophysiological signals, offers insight into the psychological mechanism, and predicts psychological variables, performance, and behaviour. Objective neurotechnology-enabled experimental methods might be used to study the behavioural and psychophysiological mechanisms and the mechanisms through which external environmental factors affect psychological variables, such as the influence of organizational factors and situational factors. Technology-enabled experimental research may well be one of the main research methods in current and future social and management research. By following the proposed experimental research design frameworks, researchers will be able to design and implement technology-enabled experimental research, and discover human psychological and behavioural mechanisms. We suggest that readers pay attention and attempt to apply this emerging experimental research methods.

Review Questions and Exercises

1 What are the advantages of using emerging technology in research?

2 What are the drawbacks of using emerging technology in research?

3 What are the processes for using emerging technology for data collection and analysis in research?

4 What ethical issues should be considered when using emerging technology in research, and how might these be addressed?

References

Armstrong, T. and Olatunji, B.O. (2012). Eye tracking of attention in the affective disorders: a meta-analytic review and synthesis. *Clinical Psychology Review* 32: 704–723.

Biasiucci, A., Franceschiello, B., and Murray, M.M. (2019). Electroencephalography. *Current Biology* 29: R80–R85.

Brouder, P. (2014). Evolutionary economic geography: a new path for tourism studies? *Tourism Geographies* 16: 2–7.

Buzsaki, G. (2006). *Rhythms of the Brain*. Oxford University Press.

Chanel, G., Rebetez, C., Bétrancourt, M., and Pun, T. (2011). Emotion assessment from physiological signals for adaptation of game difficulty. *IEEE Transactions on Systems, Man, and Cybernetics-Part A: Systems and Humans* 41: 1052–1063.

Comu, S., Kazar, G., and Marwa, Z. (2021). Evaluating the attitudes of different trainee groups towards eye tracking enhanced safety training methods. *Advanced Engineering Informatics* 49: 101353.

Dane, F.C. (1990). *Research Methods*. Brooks/Cole Publishing.

Erkan, İ. (2018). Examining wayfinding behaviours in architectural spaces using brain imaging with electroencephalography (EEG). *Architectural Science Review* 61: 410–428.

Guo, H., Yu, Y., Xiang, T. et al. (2017). The availability of wearable-device-based physical data for the measurement of construction workers' psychological status on site: from the perspective of safety management. *Automation in Construction* 82: 207–217.

Hasanzadeh, S., Esmaeili, B., and Dodd, M.D. (2017). Measuring the impacts of safety knowledge on construction workers' attentional allocation and hazard detection using remote eye-tracking technology. *Journal of Management in Engineering* 33: 04017024.

He, M., Lian, Z., and Chen, P. (2019). Evaluation on the performance of quilts based on young people's sleep quality and thermal comfort in winter. *Energy and Buildings* 183: 174–183.

Henneman, E.A., Marquard, J.L., Fisher, D.L., and Gawlinski, A. (2017). Eye tracking: a novel approach for evaluating and improving the safety of healthcare processes in the simulated setting. *Simulation in Healthcare* 12: 51–56.

Hetling, J. (2008). Comment on what is neural engineering? *Journal of Neural Engineering* 5 (3): 360–361.

Hwang, S., Jebelli, H., Choi, B. et al. (2018). Measuring workers' emotional state during construction tasks using wearable EEG. *Journal of Construction Engineering and Management* 144 (7): 04018050.

Hyönä, J. (2010). The use of eye movements in the study of multimedia learning. *Learning and Instruction* 20: 172–176.

Jebelli, H., Hwang, S., and Lee, S. (2018). EEG-based workers' stress recognition at construction sites. *Automation in Construction* 93: 315–324.

Jeelani, I., Han, K., and Albert, A. (2018). Automating and scaling personalized safety training using eye-tracking data. *Automation in Construction* 93: 63–77.

Jobin, A., Ienca, M., and Vayena, E. (2019). The global landscape of AI ethics guidelines'. *Nature Machine Intelligence* 1: 389–399.

Ke, J., Zhang, M., Luo, X., and Chen, J. (2021). Monitoring distraction of construction workers caused by noise using a wearable Electroencephalography (EEG) device. *Automation in Construction* 125: 103598.

Kim, K.H., Bang, S.W., and Kim, S.R. (2004). Emotion recognition system using short-term monitoring of physiological signals. *Medical and Biological Engineering and Computing* 42: 419–427.

Kirstein, C. (2007). Sleeping and dreaming. In: *xPharm: The Comprehensive Pharmacology Reference* (eds. D.B. Bylund and S.J. Enna) 1–4. Elsevier.

Levin, K.H. and Chauvel, P. (2019). *Clinical Neurophysiology: Basis and Technical Aspects: Handbook of Clinical Neurology Series*. Elsevier.

Li, J., Li, H., Umer, W. et al. (2020). Identification and classification of construction equipment operators' mental fatigue using wearable eye-tracking technology. *Automation in Construction* 109: 103000.

Li, J., Li, H., Wang, H. et al. (2019). Evaluating the impact of mental fatigue on construction equipment operators' ability to detect hazards using wearable eye-tracking technology. *Automation in Construction* 105: 102835.

Nilsson, D.E. (1989). Vision optics and evolution. *Bioscience* 39: 298–308.

Pfurtscheller, G. and Da Silva, F.L. (1999). Event-related EEG/MEG synchronization and desynchronization: basic principles. *Clinical Neurophysiology* 110: 1842–1857.

Rosch, J.L. and Vogel-Walcutt, J.J. (2013). A review of eye-tracking applications as tools for training. *Cognition, Technology & Work* 15: 313–327.

Salkind, N.J. (2011). *Exploring Research*, 8e. Pearson.

Teplan, M. (2002). Fundamentals of EEG measurement. *Measurement Science Review* 2: 1–11.

Wang, D., Chen, J., Zhao, D. et al. (2017). Monitoring workers' attention and vigilance in construction activities through a wireless and wearable electroencephalography system. *Automation in Construction* 82: 122–137.

Wang, D., Li, H., and Chen, J. (2019). Detecting and measuring construction workers vigilance through hybrid kinematic-EEG signals. *Automation in Construction* 100: 11–23.

Wu, Y., Kihara, K., Takeda, Y. et al. (2021). Eye movements predict driver reaction time to takeover request in automated driving: a real-vehicle study. *Transportation Research Part F: Traffic Psychology and Behaviour* 81: 355–363.

Xing, X., Li, H., Li, J. et al. (2019). A multicomponent and neurophysiological intervention for the emotional and mental states of high-altitude construction workers. *Automation in Construction* 105: 102836.

Yarbus, A.L. (1967). Eye movements during perception of complex objects. In: *Eye Movements and Vision* (ed. Y.L. Yarbus), 171–211.Boston, MA: Springer

7

Data-Driven Research

7.1 Introduction

With the advancement of information and communication technology, sensing technology and data science, more and more data are being collected for research purposes. As such, data-driven research is becoming mainstream across disciplines. While there is writing on data, database, and data science in general, this chapter focuses on data and data science as a means for research, with worked examples throughout.

We start by exploring the basic meanings of data, big data, and data science, including the features of big data and the types and characteristics of data for research. We then discuss the functions and processes of using data as a means for research and discuss technologies that can be used for generating, collecting, handling, and analysing data. We also present methods for automatically collecting data from public open access websites. We then discuss the relationship between data-driven research, qualitative research, and quantitative research.

A section focusing on visualization of the analysis results and a methodological framework is also included. In addition, we discuss the challenges of data-driven research followed by the ethical issues in data-driven research and a future outlook.

7.2 Data, Big Data, and Data Science

Data are the result of fact or observation; they are meant as a logical summary of something objective, the raw, unprocessed material used to represent something objective (Dhar 2013). Data can be continuous values, such as sounds and images (analogue data) or discrete, such as symbols and text(digital data) (Loukides 2011). Data has permeated every industry and business function, become an important factor of production, and heralded a new wave of productivity growth.

We are now in the big data era. Big data refers to the dataset that is so large that it cannot be captured, managed, processed, and organized in a reasonable amount of time by mainstream software tools, and is crucial in helping researchers conduct research (Waller and Fawcett 2013). Remote sensing, the Internet of Things (IoTs), mobile technologies, social media, and high-performance computing all relate to the emergence of big data (Hurwitz et al. 2013). Big data methods have the potential to analyse an entire population.

Research Methodology and Strategy: Theory and Practice, First Edition. Patrick X.W. Zou and Xiaoxiao Xu.
© 2023 John Wiley & Sons Ltd. Published 2023 by John Wiley & Sons Ltd.

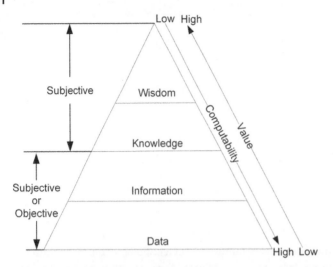

Figure 7.1 Relationships between data, information, knowledge, and wisdom.

Big data and data science are new tools for developing theory. Given the differences from existing data collection and analytical techniques to which scholars are socialized, it will take effort and skill to adapt this new practice (Grover and Kar 2017). The current model of research is post hoc analysis, where scholars analyse data collected after the temporal occurrence of the event – a manuscript is drafted months or years after the original data are collected. Therefore, the velocity or real-time application importance for practice is not a critical concern for scholars in the current research paradigm. This is a situation that can be changed with data science.

There is a sequential relationship between data, information, knowledge, and wisdom, representing the process of human cognitive transformation (Figure 7.1). Data can be seen as raw, unprocessed information. When we analyse large amounts of data, we can extract information from it to help us make decisions. After having a large amount of information, people can further summarize and synthesize it, which then forms knowledge. With a great deal of knowledge and a wealth of practical experience, one develops a view about the world, and this is when wisdom is formed.

Data-driven research involves the process of moving from data to information to knowledge, as shown in Figure 7.2. Under data-driven research, disciplines are no longer independent from each other (Huang et al. 2018). The large amounts of available data and advanced analytical techniques have combined to trigger the formation of a new interdisciplinary research area. Researchers combine data science with the characteristics and needs of their research area to achieve a process of transformation from data to knowledge. Meanwhile, researchers can improve data-driven research based on the problems encountered in the process of transforming data to knowledge.

7.3 Methodological Framework

The technology- and data-driven methodological framework is visualized in Figure 7.3. According to the objectives and focuses of the research and types of the research problems,

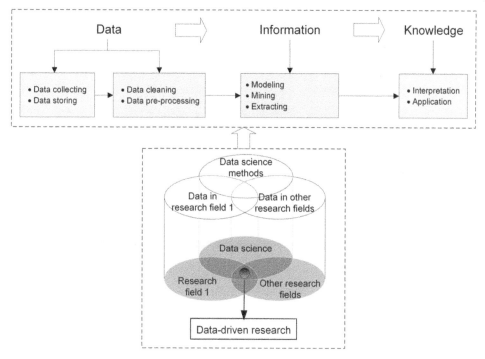

Figure 7.2 Data-driven research framework.

research aims and questions are defined and described in detail. Thereafter, either qualitative, quantitative, or mixed methods research design may be selected. Based on the chosen methods, technology-based data collection methods can be designed and implemented. After the data collection is completed, data-driven research methods can be used to analyse the collected data.

7.3.1 Data Collection

Data is the basic and core element for any research. Effective access to useful and sufficient data has always been a major issue for researchers. Data collection methods include but are not limited to: sensors, experiments, surveys, and web crawlers. Sensor-based approaches allow for continuously collecting large amount of data. For example, by getting workers to wear activity-tracking wristbands, data about their movement, heart rate, and calories burned can be gathered. Furthermore, wearable sensors can be used to collect data on moving objects' physical proximity. This data gathering approach is relatively nonobtrusive and provides information about employees in natural settings. Sensors can also be used to monitor the work environment, e.g., temperature, humidity, light, and noise, or movements and gas consumption of vehicles (George et al. 2016).

Survey is a data collection method that gathers the thought, feeling, and emotion of certain groups of people regarding particular topics. It includes but is not limited to questionnaires, document and archival records collection, and observation, and can involve data companies.

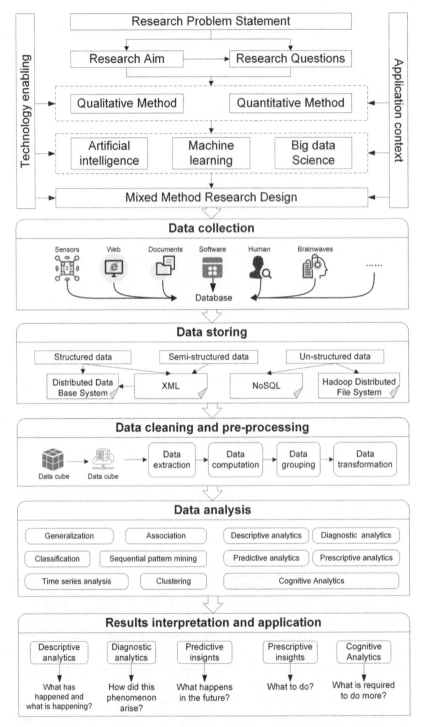

Figure 7.3 Technology- and data science-enabled research design and process.

Web crawlers allow for the automated extraction of large amounts of data from websites. Web crawler programs are widely available and often free, in the form of plugins for popular web browsers such as Google Chrome. Some websites, e.g., Twitter, offer application program interfaces (APIs) to ease and streamline access to the contents on their websites. Web crawlers can be programmed to extract numeric data, textual data, audio data, and video data. A list of sources and methods for obtaining research data is presented in Table 7.1.

Table 7.1 Sources and methods for obtaining research data.

Sources and methods	Descriptions	Data types
Publicly open data		
Statistical Yearbook	The statistical yearbook is a comprehensive, systematic, and continuous record of annual economic and social developments through highly dense statistical data, mainly in the form of statistical charts and analytical notes.	Quantitative
Open access industry record	Data provided to the public by companies that are relevant to the business and do not involve business confidential.	Qualitative or quantitative
Government public data	The government makes its data and information resources available to the public in a manner that meets open data requirements.	Qualitative or quantitative
Confidential data		
Government confidential data	The use of data is subject to participation in government projects and the signing of confidentiality agreements.	Qualitative or quantitative
Industry confidential data	The use of the data is subject to a partnership with a company and a confidentiality agreement.	Qualitative or quantitative
Data company	Data companies provide data in accordance with the needs of the research.	Qualitative or quantitative
Experiment-based data		
Natural experiment	The use of instruments and equipment to change, control, or simulate the object of research to make something (or a process) happen or to reproduce, in order to understand natural phenomena, nature, and natural laws, while eliminating external influences and highlighting the main factors.	Quantitative

(Continued)

Table 7.1 (Continued)

Sources and methods	Descriptions	Data types
Social experiment	A type of research done in fields like psychology, economics, or sociology to see how people behave in certain situations or how they respond to particular policies.	Qualitative or quantitative
Survey-based data		
Interview	Interview is a qualitative research method to generate in-depth and rich data, e.g., behaviour, attitude, beliefs, norms and values.	Qualitative
Questionnaire	The questionnaire systematically documents the survey in the form of questions.	Qualitative or quantitative
Document and archival records	Provision of information, evidence or written, printed or electronic material as an official record.	Qualitative or quantitative
Direct observation	Researchers view and record what is happening or what a person is doing.	Qualitative or quantitative
Participant observation	Researchers take on different specific roles and are actually involved in the behaviour under research, and record what is happening.	Qualitative or quantitative
Technology-enabled data		
Web crawler	Web crawler technology is based on certain strategies written by programmers to automatically collect information from the Internet.	Qualitative or quantitative
Sensor	A sensor is a detection device that senses the measured information and can transform the sensed information into an electrical signal or other required forms of information output according to certain rules to meet the requirements of information transmission, processing, storage, display, recording and control.	Quantitative
Wi-Fi	Wi-Fi is an efficient and cost effective method of collecting data on the position and number of people.	Quantitative
Meter	A device that measures the amount of gas, water, or electricity used.	Quantitative

7.3.2 Data Storage

Data storage depends on the size of the data. For relatively small datasets, no dedicated storage solutions are required. Microsoft Excel, SAS, SPSS, and Stata can hold data for many subjects and variables; Excel can hold 1,048,576 rows and 16,384 columns. Big data typically requires a large storage capacity. If the data storage capacity needed exceeds the individual computers' storage capacity, data can be stored in a cloud drive. Larger datasets can be stored in a relational database using structured query language (SQL). Relational databases store data in multiple tables that can be easily linked with each other, and there are open source relational databases available, e.g., MySQL and PostgreSQL (George et al. 2016).

Data can be classified into structured, unstructured, and semi-structured types. The distributed database system (DDBS) can be used to store and manage structured data, the XML can be used to describe semi-structured data that are unified into a standardized data format and stored in the relational database management system and distributed database management systems (such as the Hadoop distributed file system, HDFS), and the unstructured query language (NoSQL) manages and stores the unstructured data (Huang et al. 2018).

7.3.3 Data Cleaning and Preprocessing

Some data collected may not meet the requirements, requiring cleaning. Data cleaning handles duplicate data, missing data and outliers as follows:

1) Duplicate data processing: duplicate data may influence the results of statistical analyses and they should be removed. The standard method of detecting duplicate data is to sort all the data and then to check for and delete any identical consecutive records (Rahm and Do 2000).
2) Dealing with missing data: missing data may occur for many reasons, such as participant reluctance or forgetfulness in responding to questions, malfunctioning data collection equipment, and data entry errors (Osborne 2013). There are two ways for researchers to handle missing data. The first is to omit elements from a dataset that contain missing values, whilst the second is to replace missing data by using global constant, moving average, imputation or inference-based models (Hastie et al. 2009; Xiao and Fan 2014).
3) Outlier detection: an outlier is an observation which deviates markedly from other observations (Barnett and Lewis 1994). Outliers do not always equal errors, which should be detected but not necessarily removed (Jonge and Loo 2013). Outliers are identifiable using unsupervised clustering, semi-supervised recognition and supervised classification methods (Maimon and Rokach 2010).

Data preprocessing follows a number of steps: data extraction, the selection of parts of the data according to the needs of the research; data computation, where various arithmetic and logical operations help obtain further information; efficient grouping of relevant data; normalization or naturalization of data to suit the needs of data analysis algorithms through z-score normalization, maximum–minimum normalization, normalization by decimal scale and other avenues.

7.3.4 Data Analysis

The data collected may contain many variables available for theory development. It is common to observe hundreds or even thousands of variables. In theory testing, the variables of interest, and possible control variables, are known. In exploratory studies, where the aim is to develop a new theory based on the data, we need statistical methods to assist in variable selection. Researchers might try to estimate models using all possible combinations of explanatory variables and then compare model fit using a likelihood-based criterion. However, with the increasing of the number of explanatory variables, the number of combinations increases exponentially, rendering this approach unfeasible. In a situation where there are many explanatory variables, ridge and lasso regression, principal components regression, partial least squares, Bayesian variable selection, and regression trees are all designed to assist with data analysis (Varian 2014). Some methods of data analytics are explained in Table 7.2.

Table 7.2 Data analytics methods.

Analytic types	Explanations
Generalization	The process of abstracting a large amount of data in a database that is relevant to the research objectives.
Association	Discovers interdependency or relationships hidden in large datasets. Based on different rules, association can be divided into simple association, timing association and causal association.
Clustering	Aims to organize data into categories on the basis of their similarities.
Classification	Constructs a classification function according to the characteristics of the dataset, and identifies and assigns categories to a collection of data to allow for more accurate analysis.
Sequential pattern mining	Discovers high-frequency, time-related or other sequence-related subsequences from the sequence database.
Time series analysis	Mainly used to summarize the data sequence or trend and to monitor the periodic change of data. It extracts information from a large number of time series data with one or more time attributes that are not known to people in advance but can be potentially useful for predictions (Huang et al. 2018).
Descriptive analytics	A statistical method used to interpret data and to provide insight into patterns and meaning.
Diagnostic analytics	The process of using data to determine the causes of trends and the interdependence between variables. This can be seen as a logical next step after using descriptive analysis to identify trends.
Predictive analytics	Designed to analyse data, discover patterns, observe trends, and use that data to forecast the future.
Prescriptive analytics	Used to make decisions through the analysis of raw data, which is the following step after the predictive analytics.
Cognitive analytics	Combines a number of intelligent technologies, e.g., artificial intelligence, deep learning, machine learning, and reinforcement learning, to learn from large datasets and give structure to the unstructured data.

Common software for data analysis includes Microsoft Excel, R, and Python. With a wide range of powerful functions, such as table creation, statistical analysis, and matrix calculations, Excel covers the basic functions. However, in today's era of big data and artificial intelligence, Excel is a customized software that cannot be freely modified and is not up to handling large volumes of data. Excel is more suitable for descriptive analysis, such as comparative analysis, trend analysis, and structural analysis. In comparison, Python and R, are powerful and flexible and can be written in programming codes to perform any operation required.

7.3.5 Visualization and Reporting

1) Visualization

Data visualization aims to communicate information clearly and effectively through graphical means, helping researchers to quickly understand information, analyse trends in data, and identify patterns in relationships. The quality of information is largely determined by the way in which it is presented, and data visualization is a good way of presenting data, combining data analysis techniques with graphical techniques to interpret and communicate the results of analysis clearly and effectively (George et al. 2016). With the rise of data science, many new visualization tools (e.g., Python, R, and Tableau) allow for easy application of multiple selections and visualization for large datasets.

Figure 7.4 illustrates the important role of data visualization in research. Specifically, visualization designers turn cluttered data into intuitive visualization with the help of algorithms. Visualization transforms data into information, and allow users to form conceptual models. These conceptual models in turn produce visualizations that correspond to different scenarios (McBrien et al., 2018).

2) Reporting

When it comes to reporting data, analytical results, and research findings, one of the key challenges is completeness. The variety of big data makes it important to clearly describe

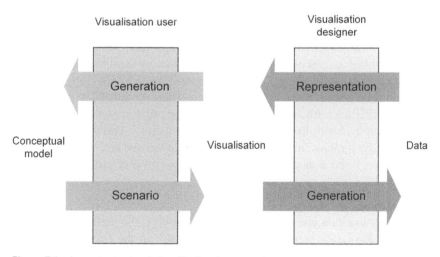

Figure 7.4 Important role of visualization in research.

the different data sources. Steps taken in preprocessing and merging of the data should be carefully described. It is also important to explain the context in which data is gathered and presented. Readers should understand the decisions that have been made in preparing the data for analysis (e.g., Haas et al. 2015). For example, the words 'cost' and 'liability' may express negative sentiment in some settings, whereas they are more neutrally used in financial texts. It should be made clear to the readers whether weights are applied, and which other decisions have been made in preparing.

7.4 Examples of Data-driven Research

The following sections present an example 'Discovery of new safety knowledge from mining large injury dataset in construction projects' (Xu and Zou 2021), a research conducted by the authors. The research aimed to discover the characteristics, patterns, mechanisms, and knowledge from mining a large injury dataset, and to develop strategies for mitigating safety risks and reducing accidents so as to improve construction safety performance. For more details, seek the article published in *Safety Science* (https://doi.org/10.1016/j.ssci.2021.105481).

1) Description of the dataset

The data used in this study were provided by well-established construction companies operating nationwide and focused on all types of construction contracting and service providers. As such, the data are highly representative. The dataset contains three categories of information, namely basic information of time and the injured person, injury detail description information, and reason (causes) of injury or how the injury happened. The basic structure of the dataset is as shown in Figure 7.5. From the basic information, we can know the age, gender, and occupation of the injured person, as well as the exact time of the injury. For example, the injury numbered CDY-0010565 occurred at 11:40 pm on 3 April, and the injured worker is a 25-year-old male concreter. The injury information includes the nature of injury (e.g., superficial injury, muscle/tendon strain, contusion), bodily location (e.g., wrist, back, hand, leg, and ankle), and detailed description of the injury. In addition, companies conducted in-depth investigations of the causes of every injury. The content and report of investigation include event description and mechanism of injury (e.g., hitting moving objects, muscular stress while lifting, carrying or putting down objects, and hitting stationary objects).

Data types included nominal data, text data, and interval data. Nominal data uses a nominal scale to label variables without any quantitative value, such as gender, occupation category, nature of injury, bodily location, and mechanism of injury. Text data usually consists of sentences or paragraphs. In this study, event description and detailed descriptions of injury are text data. Interval data represent the numerical value of variables, e.g., date, time and age.

2) Data preparation

Before data analysis and data visualization, data preparation is required to ensure data quality. Three rounds of data preparation were carried out. In the first round of data

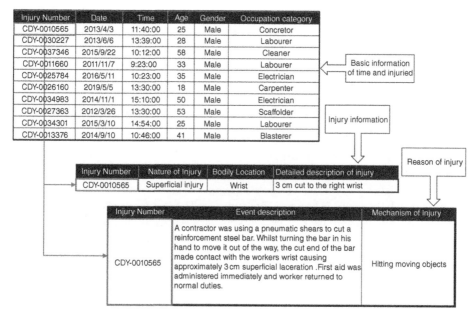

Figure 7.5 Database for construction injury details.

preparation, all unrelated data are deleted. This research focuses on construction injury; thus, non-injury-related data, such as estimated cost of damage and name of responsible person, are deleted. In the second round of data preparation, all data with missing values are removed. In the third round all data are checked to ensure there is no error data and outliners.

3) Data analysis methods and processes

Statistical analysis: in this example, descriptive statistics analysis, probability distribution analysis, and linear regression analysis are applied. Descriptive statistics could describe the basic feature of the data and this study uses descriptive statistics to describe the injury proportions of body parts. Probability distribution analysis is a method that allows us to create a statistical distribution with a specific pattern to describe the probability of injury. Moreover, from probability distribution, we can estimate the probability when (time) an injury event might be likely to occur. Linear regression is applied to analyse variation over time in the proportion of injuries due to different natures of injury. All data analysis were performed using R, a free software environment for statistical computing and graphics.

Data visualization: data visualization methods are used in this example to visualise the results of the statistical analysis. First, probability distribution of safety incidents by month and time of the day are presented, from which months and times of day with a higher frequency of safety incidents can be easily identified. Second, bodily locations of injury and corresponding injury proportions are shown in the form of body structure diagrams. Third, age distributions of the injured, as well as bodily locations of injury, are presented. Forth, the relationship between injured body parts and mechanism of injury, and the relationship between injured body parts and nature of injury are presented in the form of scatterplots with smoothed densities. All data visualization was performed using R, a free software environment for statistical computing and graphics.

Association rule mining (ARM): this example employs ARM to discover the rules between occupation category, nature of injury, bodily location, and mechanism of injury. Apriori algorithm is selected to perform ARM as it is for frequent item set mining (Agarwal and Srikant 1994). Rules with a support above 0.006 and confidence above 0.25 are first selected (Verma et al. 2014). After that, rules are checked to ensure that whether they make sense. Unreasonable rules are eventually removed.

4) Analysing results of time of injury

From the perspective of calendar months of the injury events, February, October, and November had higher percentages of injuries (Figure 7.6). The research data were in the southern hemisphere where February is hot summer. The temperatures can reach to the high 30 degrees and become heatwaves. Workers wear less clothing during the summer months, making them more susceptible to cuts, bumps, and bruises. Compared with summer, injury incidents are lower in winter (June, July, and August). On one hand, the winter is relatively mild. This provides a more suitable environment for construction. Moreover, workers wear more clothing in winter, which also protects them from injury to some extent.

December and January are found to have a lower percentage of injuries, despite being summertime. The main reason may be that construction work is least intensive during this two-month period due to Christmas and New Year holidays. Similarly, construction work is less in April as most workers take the Easter holiday period off. It is interesting to see that incidents increased rapidly after every holiday. This may mean that workers need a period of adjustment after the holidays to adjust themselves to the pace of work. Otherwise, workers are likely to be injured as they are not comfortable with this shift from vacation to work.

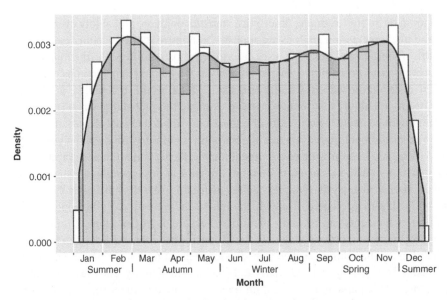

Figure 7.6 Probability distribution of safety incidents by calendar month.

Construction workers usually start work at 7:00 or 7:30 in the morning and finish at 15:30 or 16:00 in the afternoon Monday to Friday, and Saturday mornings. As shown in Figure 7.7, the period between 7:00 a.m. and 9:00 a.m. witnessed a dramatic increase in the probability of injury event. The reason may be that workers' minds were not on work at the beginning of the day. It is also found that most injury events occurred between 11:00 a.m. and 12:00 noon. This was arguably due to the fact that fatigue contributes to injury events.

After a lunch break, however, there was a small increase in the probability of injury events. The same phenomenon also occurred after morning tea. On one hand, a short break could relieve physical and mental fatigue of workers; on the other hand, these breaks may affect working state.

Figure 7.7 also shows that not many injuries occurred before 7:00 and after 17:00. Construction work is not normally allowed to be carried out during these time periods, unless it is essential to undertake a construction task outside normal working hours (such as on-road/sea transport of construction materials or equipment). Furthermore, if work is carried out outside the normal 7:30–17:00 daily routine, the costs for worker's wages would be doubled, which may significantly increase project costs.

5) Analysing results of bodily locations, organs, and systems of injuries

The proportion of injured body parts, organs, and systems is presented in Figure 7.8. It can be seen that the hand has the highest percentage of injuries (27.75%), followed by the back (11.59%), eyes (10.39%), and feet (6.72%). If extending the hand to include the arm, wrist, and elbow, the percentage is 36.54%, which is over one-third of the injuries. Similarly for feet, if extended to include leg, knee, and hip, the percentage is as high as 19.95%. Based on the event descriptions, the main causes of hand injuries include 'being hit by moving/falling objects', 'hitting stationary objects', 'being trapped between stationary and moving

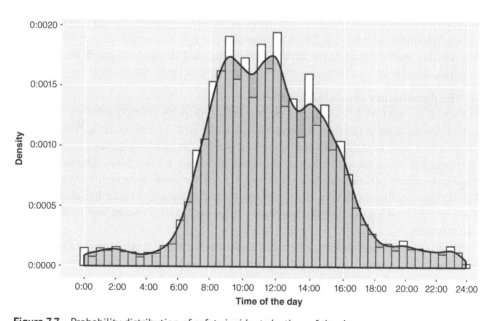

Figure 7.7 Probability distribution of safety incidents by time of the day.

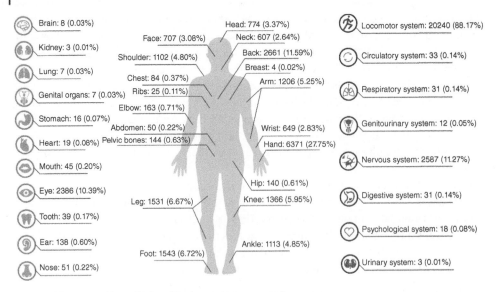

Brain: 8 (0.03%)

Kidney: 3 (0.01%)

Lung: 7 (0.03%)

Genital organs: 7 (0.03%)

Stomach: 16 (0.07%)

Heart: 19 (0.08%)

Mouth: 45 (0.20%)

Eye: 2386 (10.39%)

Tooth: 39 (0.17%)

Ear: 138 (0.60%)

Nose: 51 (0.22%)

Face: 707 (3.08%)

Shoulder: 1102 (4.80%)

Chest: 84 (0.37%) Ribs: 25 (0.11%)

Elbow: 163 (0.71%)

Abdomen: 50 (0.22%)

Heart: 19 (0.08%) Pelvic bones: 144 (0.63%)

Leg: 1531 (6.67%)

Foot: 1543 (6.72%)

Head: 774 (3.37%)

Neck: 607 (2.64%)

Back: 2661 (11.59%)

Breast: 4 (0.02%)

Arm: 1206 (5.25%)

Wrist: 649 (2.83%)

Hand: 6371 (27.75%)

Hip: 140 (0.61%)

Knee: 1366 (5.95%)

Ankle: 1113 (4.85%)

Locomotor system: 20240 (88.17%)

Circulatory system: 33 (0.14%)

Respiratory system: 31 (0.14%)

Genitourinary system: 12 (0.05%)

Nervous system: 2587 (11.27%)

Digestive system: 31 (0.14%)

Psychological system: 18 (0.08%)

Urinary system: 3 (0.01%)

Figure 7.8 Proportion of injured body parts, organs, and systems.

objects', and 'contact with electricity'. The very high rate of 27.75% of hand injuries indicates the need for wearing gloves at construction work sites. Although some construction companies have introduced 'wearing gloves' as a mandatory requirement, most companies are still not making this mandatory. This is a global problem that deserves more attention of policymakers, construction managers, and researchers.

Regarding the injured systems, as shown on the right-hand column of Figure 7.8, the locomotor system has the highest percentage of injuries, at more than 88%; the nervous system accounted for over 11%. These two systems together accounted for over 99% of injuries. The locomotor system is the main body system used by workers during construction activities. It generally interacts or comes into direct contact with the external environment.

6) Age distribution of injured workers

The age of the workers who sustained injuries ranged from 18 to 70 (Figure 7.9). Figure 7.9 shows that workers under and around the age of 20 had low injury rates. This may be because they were being supervised by a senior worker (particularly if they are an apprentice) hence more cautious. However, the trend increased rapidly from the mid-20s and those particularly between the ages of 25 and 30 are most likely to be injured (Figure 7.9). Furthermore, younger workers have not experienced many safety incidents, and they are more likely to take risks and disobey safety rules and expose themselves to dangerous environments.

The relationships between age and bodily locations of injury are presented in Figure 7.10. Nine bodily locations are considered, i.e., ankle, arm, back, eye, foot, hand, knee, leg, and shoulder, as they have a higher percentage of injuries. In Figure 7.10, each 'violin' plot represents the probability distribution of age by a bodily location of injury. The white dot

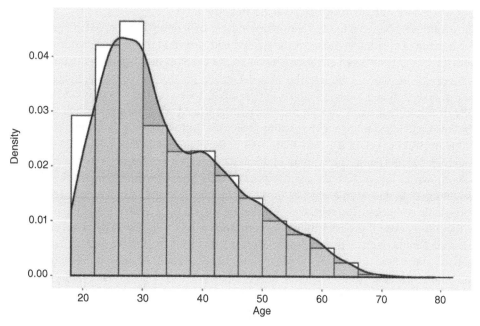

Figure 7.9 Age distribution of the injured.

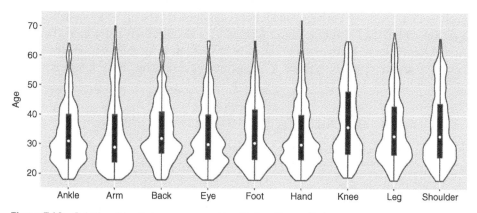

Figure 7.10 Relationships between age and bodily locations of injury.

represents the median age. For example, the white dot of ankle indicates that the median age of workers with ankle injuries is 31. The thick black bar in the centre shows the inter-quartile range. From Figure 7.10, it can be found that the presented bodily locations of injury were mainly on young workers under 30 years. Meanwhile, older workers are more likely to have injuries to the knee, shoulder, leg, and back. This may be due to the fact that the functions of these body parts have declined as workers get older. Therefore, older workers should be adequately protected in these body parts.

7) Causes and nature of injury relating to body parts

The relationship between injured body parts and causes of injury is presented in Figure 7.11. The horizontal coordinate indicates the injured body part and the vertical coordinate indicates the cause/mechanism of injury. The colour indicates the number of overlapping points in the graph: the darker colour of the area, the greater the number of overlapping points of injured body part and causes of injury. For example, *Area a* contains hand injuries; specifically, it shows that hand injuries are strongly associated with hitting moving/stationary objectives and being hit by moving/falling objectives. Although the number of overlapping points in *Areas b*, *c*, and *d* is less than *Area a*, they still indicate possible links between injured body parts and mechanism of injury and cannot be overlooked.

As shown in Figure 7.11 the main causes of body injury include being hit by objects, weight bearing, and hitting objects. More effort should be paid to eliminate, reduce, or mitigate these causes. However, whenever an injury has happened, it could happen again in the future, even if the probability of occurrence is relatively low. Therefore, sufficient attention should be given to every cause of injury. It is also found that exposure to environmental cold or heat or a traumatic event is less likely to result in injury to body parts. On one hand, these sources of danger were less likely to occur. On the other hand, workers were better protected against these sources of danger.

The relationships between injured body parts and nature of injury is shown in Figure 7.12. The horizontal coordinate indicates the injured body part, and the vertical coordinate indicates the nature of injury. *Area a* indicates the most likely nature of injury to hand, back, and eye. Especially, hands are more likely to have a superficial injury and laceration or open wound. The main reason includes but is not limited to being hit by an object, contact with electricity, and hitting an object. Compared with other body parts, hands have the most diverse nature of injury. Due to the large amount of work that requires the use of hands, they are more easily exposed to dangerous environment. Back injury is highly

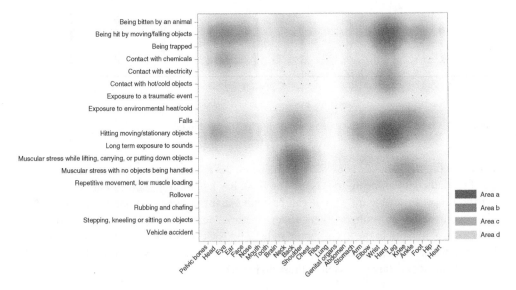

Figure 7.11 Relationship between injured body parts and mechanism (causes) of injury.

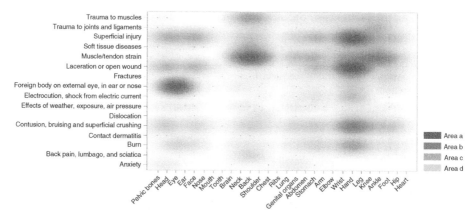

Figure 7.12 Relationship between the injured body parts and nature of injury.

related to muscle/tendon strain. If muscle/tendon strain is not treated promptly, workers may experience recurring injuries or pain as well as weakness in the muscle at work. Eyes are more likely to suffer from 'foreign body on external eye', which is usually caused by being hit by objects or exposure.

8) Results of association rule mining (ARM) analysis
In total 128 rules were derived from the datasets based on association rule mining methods. These 128 rules describe in detail the possible injuries to different body parts as well as the corresponding causes, providing a basis for workers and management to develop targeted protective measures. The top 20 rules are summarized in Table 7.3, which shows not only the relationship between the injury and body parts, but also the causes/mechanism of the injuries. Most of the rules may seem to be common sense and can be obtained from work experience and domain knowledge. However, they reflect that these rules are common on construction sites and cannot be ignored. For instance, Rule 1 in Table 7.3 states that, if workers' hand is in contact with electricity, they could have an electric shock from the current. This can be easily interpreted, as construction sites are prone to electrical wiring leaks and workers mainly use their hands to operate equipment at construction sites. As another example, Rule 8 describes that workers' eyes are more likely to sustain 'foreign body on external eye'. This can also be easily understood, as there is dust on the construction site that could easily get into the eyes of workers who are not wearing protective glasses. Rule 19 indicates that the face may be hit by moving objects. Accordingly, it is suggested that workers wear face protection on construction sites where there are moving objects.

7.5 Ethics in Data-driven Research

Data science development and application must be done with a mind to ethics, from programming, which requires ethical data collection and must not absorb pre-existing social prejudices, to the extent of how much the results should be relied upon or trusted. There are certain ethical risks unique to data-driven research, such as technology uncertainty,

Table 7.3 The top 20 results of association rule mining.

No.	Rules	Support(%)	Confidence(%)	Lift
1	Hand, Contact with electricity => Electric shock from current	0.80	87.57	78.91
2	Contact with electricity => Electric shock from current	1.08	85.37	76.92
3	Contact with hot objects => Burn	2.70	93.23	23.92
4	Tradesmen, Eye => Exposure to other and unspecified environmental conditions	2.95	60.21	12.09
5	Tradesmen, Eye, Exposure to other and unspecified environmental conditions => Foreign body on external eye,	2.85	96.50	10.42
6	Tradesmen, Eye => Foreign body on external eye	4.39	89.47	9.66
7	Exposure to other and unspecified environmental conditions => Foreign body on external eye in ear or nose or in respiratory digestive or reproductive tract	4.35	87.25	9.42
8	Eye => Foreign body on external eye in ear or nose or in respiratory digestive or reproductive tract	8.40	85.91	9.28
9	Eye => Exposure to other and unspecified environmental conditions	4.48	45.80	9.20
10	Eye, Single contact with chemical or substance => Foreign body on external eye	1.12	72.58	7.84
11	Ankle => Stepping, kneeling, or sitting on objects	2.46	47.41	7.44
12	Single contact with chemical or substance => Burn	0.77	26.28	6.74
13	Tradesmen, Knee => Stepping, kneeling, or sitting on objects	0.73	29.68	4.66
14	Back => Muscular stress while lifting carrying or putting down objects	4.28	34.93	4.51
15	Tradesmen, Back, Muscular stress while handling objects other than lifting carrying or putting down => Muscle/tendon strain (nontraumatic)	1.34	99.24	4.37
16	Tradesmen, Muscular stress while lifting carrying or putting down objects => Muscle/tendon strain (nontraumatic)	2.66	93.45	4.11
17	Tradesmen, Back => Muscle/tendon strain (nontraumatic)	4.25	90.85	4.00
18	Hand, Hitting stationary objects => Laceration or open wound not involving traumatic amputation	4.34	43.19	2.89
19	Face => Being hit by moving objects	1.49	46.45	2.85
20	Hand => Laceration or open wound not involving traumatic amputation	8.84	32.92	2.20

Note: Lift is a measure of 'interestingness' of a discovered rule.

human-centrism, technology overuse, data privacy risk, and gender bias risk. All data collection processes must be 'ethical, secure, and transparent' (Rouhiainen 2019). Questions to ask about ethics include but not limited to:

1) What do we know about the society and why?
2) How do we know what we know about the society?
3) Data always reflect the biases and interests of those doing the collecting, how can we minimize this bias?
4) Who is designing a technological intervention for a social setting, who participates, and who is affected by it?
5) How do we assess the techno–social–cultural system interactions?
6) How do we interpret the different types of knowledge?

Jobin et al. (2019) surveyed the 84 artificial intelligence (AI) guidelines existing across the world, at local, national, and international levels, published mostly by institutions, councils, and companies. In conducting this extensive survey, they identified 11 main ethical principles: transparency, justice and fairness, nonmaleficence, responsibility, privacy, beneficence, freedom and autonomy, trust, sustainability, dignity, and solidarity. These principles are suitable for application to data-driven research. Using the 11 principles specified by Jobin et al. (2019), Table 7.4 suggests ways these ethical principles could be applied and implemented.

Table 7.4   Ethical principles for data-driven research (Adapted from Jobin et al. 2019).

Principle name	Definition (Adapted from Oxford Dictionary)	Application in data-driven research
Identity, equality, & transparency	Open to public scrutiny, easy to perceive or detect	There should be explicit communication about how data are being collected and used
Justice & fairness	The quality of being just, fair, or reasonable	Programming and algorism should avoid unwanted bias, and especially avoid replicating systems of social discrimination. Data must be accessed fairly.
Nonmaleficence	Not causing harm or destruction	The programming and algorism should prioritize user well-being above all else, and avoid causing harm, even when unintentional.
Responsibility	The state or fact of being accountable or to blame	Researchers should act with integrity with regards to big data and data queries and usage, being upfront about processes and accountable for any liabilities.
Privacy	The state of being free from public attention	Any data gathered should not be used or shared without consent.
Freedom & autonomy	Lack of external control or influence	Researchers should have a right to control all data.
Trust	The state of being responsible for someone or something	Researchers should be able to trust and work with different data stakeholders as a whole.
Beneficence	Resulting in good	Programming and algorism must specifically be designed and used for the research.

(Continued)

Table 7.4 (Continued)

Principle name	Definition (Adapted from Oxford Dictionary)	Application in data-driven research
Sustainability	Conserving an ecological balance by avoiding depletion of natural resources	The use of data should be as eco-friendly as possible.
Dignity	The state of quality of being worthy of honour or respect.	Data science (such as programming and algorism) should not diminish human dignity or overtake their roles.
Solidarity	Unity or agreement of feeling or action	Programming and algorism should be implemented with input from or consultation with all parties concerned.

Research ethics has also been discussed in Chapter 1. For general ethical principles, please refer to relevant contents in Chapter 1.

7.6 Challenges and Future Outlook

Data have become a strategic resource comparable to oil and gas. In the last two decades, the world has made significant progress in both data-related technology and policy. However, data-driven research is still being challenged in a number of ways.

1) Researchers' knowledge in data science is relatively lacking. Data-driven research requires researchers to integrate their research field knowledge with data science techniques. This requires researchers to have a certain degree of data science foundational knowledge and application skills.
2) Data barriers and data security issues. Data-driven research relies on huge amount of data, which not only need to be collected by the researchers but also to be obtained from elsewhere through technologies. Considering data security and competition issues, data owners may be unwilling to share data, which hinders data-driven research.
3) Arithmetic power and algorithms need to be further developed. Currently, there are various types of data, mostly unstructured or semi-structured data, and traditional data processing techniques and existing equipment can no longer meet the requirements of big data development.

Having discussed the above, the development of data science is bound to have a profound impact on the future of research in all disciplines. More and more data will be generated and collected, driven by technologies, including the Internet of Things, 5G and artificial intelligence (AI), and so on. The development of computer hardware and software will further enhance algorithms that are capable of analysing big data.

7.7 Summary

With the advancement of new emerging technologies and data science, more and more data are collected for research purposes. As such, data-driven research has been applied more and more in research in many disciplines. This chapter started by explaining basic concepts and methods, then provided a data-driven research methodological framework. To illustrate its applications, a worked example is presented in detail. Finally, the ethical issues and challenges are also discussed. We strongly suggest that readers pay attention to data-driven research as it is the future trend and will be applied more and more in future research.

Review Questions and Exercises

1 What do the terms big data and data science mean?
2 How should big data and data science methods be used in research?
3 What changes do big data and data science bring to research methods?
4 How will researchers use big data and data science methods in research?
5 What are the different analytic methods in data-driven research?

References

Agarwal, R. and Srikant, R. (1994) Fast algorithms for mining association rules, *Proceedings of the 20th VLDB Conference* 487–499.

Barnett, V. and Lewis, T. (1994). *Outliers in Statistical Data*, 3e. Wiley.

Dhar, V. (2013). Data science and prediction. *Communications of the ACM* 56 (12): 64–73.

George, G., Osinga, E.C., Lavie, D., and Scott, B.A. (2016). Big data and data science methods for management research: from the editors. *Academy of Management Journal* 59 (5): 1493–1507.

Grover, P. and Kar, A.K. (2017). Big data analytics: a review on theoretical contributions and tools used in literature. *Global Journal of Flexible Systems Management* 18 (3): 203–229.

Haas, M.R., Criscuolo, P., and George, G. (2015). Which problems to solve? Online knowledge sharing and attention allocation in organizations. *Academy of Management Journal* 58 (3): 680–711.

Hastie, T., Tibshirani, R., and Friedman, J. (2009). *The Elements of Statistical Learning: Data Mining, Inference and Prediction*, 2e, Springer Series in Statistics. New York: Springer.

Huang, L., Wu, C., Wang, B., and Ouyang, Q. (2018). Big-data-driven safety decision-making: a conceptual framework and its influencing factors. *Safety Science* 109: 46–56.

Hurwitz, J., Nugent, A., Halper, F., and Kaufman, M. (2013). *Big Data for Dummies*. Wiley.

Jobin, A., Ienca, M., and Vayena, E. (2019). The global landscape of AI ethics guidelines. *Nature Machine Intelligence* 1 (9): 389–399.

Jonge, E. and Loo, M. (2013). *An Introduction to Data Cleaning with R*. Heerlen, The Netherlands: Statistics Netherlands.

Loukides, M. (2011). *What is Data Science?* O'Reilly Media, Inc.

Maimon, O. and Rokach, L. (2010). *Data Mining and Knowledge Discovery Handbook*, 2e. New York: Springer.

McBrien, P. and Poulovassilis, A. (2018) Towards data visualisation based on conceptual modelling. *37th International Conference on Conceptual Modeling, Xi'an, China, 22–25 Oct 2018.*

Osborne, J.W. (2013). *Best Practices in Data Cleaning: A Complete Guide to Everything YouNeed to Do before and after Collecting Your Data*. Thousand Oaks: SAGE Publications.

Rahm, E. and Do, H.H. (2000). Data cleaning: problems and current approaches. *Bulletin of the Technical Committee on Data Engineering* 23 (4): 3–13.

Rouhiainen, L. (2019). How AI and data could personalize higher education. Cambridge, MA. *Harvard Business Review*. https://hbr.org/2019/10/how-ai-and-data-could-personalize-higher-education.

Varian, H.R. (2014). Big data: new tricks for econometrics. *Journal of Economic Perspectives* 28 (2): 3–28.

Verma, A., Khan, S.D., Maiti, J., and Krishna, O. (2014). Identifying patterns of safety related incidents in a steel plant using association rule mining of incident investigation reports. *Safety Science* 70: 89–98.

Waller, M.A. and Fawcett, S.E. (2013). Data science, predictive analytics, and big data: a revolution that will transform supply chain design and management. *Journal of Business Logistics* 34 (2): 77–84.

Xiao, F. and Fan, C. (2014). Data mining in building automation system for improving building operational performance. *Energy and Buildings* 75: 109–118.

Xu, X. and Zou, P.X.W. (2021). Discovery of new safety knowledge from mining large injury dataset in construction. *Safety Science* 144: 105481.

8

The Fifth Research Paradigm

Hybrid Natural-Social Sciences Methods Research

8.1 Introduction

This chapter focuses on how natural science and social science research methods may be hybridized to form a fifth research paradigm, a multidisciplinary or cross-disciplinary approach. As a matter of fact, and as a system of knowledge for human beings to understand the laws of movement and development of the world, natural science and social science share the general attributes of science. All sciences were once considered types of philosophy. We live in a world where intricate connections between different things are prevalent and where both natural science and social science are united in the process of human beings gaining understanding of the world. As the complexity of research problems increases, there is bound to be a great deal of crossover between natural and social sciences.

Let us think about today's complex social communities as a mega-open system. Complex system thinking and methodologies are needed to solve complex social problems, which may include human, technology, politics, economics, and culture. Given such a high level of complexity, it is unlikely that a single research method within the social research domain would be able to solve the problem. It would be useful to look into using computational and experimental approaches to solve social problems. That is, natural science methods could be applied to solve complex social problems. For example, a social network analysis method can be used to model and simulate the social interactions between different social actors in a complex social system; it would also be useful to look into the social problem from an agent-based modelling approach, which brings natural science methods into play. Some of these quantitative analytical methods have been discussed in Chapter 3.

Let us also think about how the current rapid development and application of emerging information and communication technologies (ICT) and data science may have impacted on the society, leading to social phenomena becoming more complicated and complex. With the application of ICT and data science, the data that underpins the phenomena might be automatically collected, forming a big database. This big database can be analysed using data science methods, which means that, in reality, we use natural science methods to analyse social problems.

Research Methodology and Strategy: Theory and Practice, First Edition. Patrick X.W. Zou and Xiaoxiao Xu.
© 2023 John Wiley & Sons Ltd. Published 2023 by John Wiley & Sons Ltd.

Having said this, the most important thing comes back to problem definition and research aims. Once these are properly defined and clearly described, and independent variables and dependent variables are defined together with their interrelationships, natural science research methods can be used to solve these problems, and social science methods can be used to explain these problems and the solutions proposed through the natural science research process. This means that within these processes, typical social research methods should also be used where suitable and appropriate. This indicates that a hybrid of social science methods with natural science methods becomes a new method of research design, the fifth research paradigm. The following sections provide explanation to each of these research paradigms.

8.2 The First Research Paradigm

For natural science, the first paradigm is empirical evidence-based research, which is based on experiments; it is the earliest human scientific research, characterized mainly by the recording and description of natural phenomena. For several thousand years, science has been referred to as the empirical study of natural phenomena. For example, by observing astronomy over a long period of time, the ancient Chinese summed up the 24 solar terms that guide agricultural production activities; Mozi and his students performed the world's first small-aperture imaging experiment, explaining the cause of inverted images and the first scientific explanation of the linear propagation of light. Although the first research paradigm made an important contribution to the development of science in ancient mankind, the limitations of the experimental conditions of the time made it difficult to achieve a more precise understanding of complex natural phenomena, with the focus on observation and thinking rather than explanation.

For social science, the first research paradigm is qualitative research. In the first phase of qualitative research (since the seventeenth century), mankind began to give birth to the germ of ideas about social understanding, exploring the self-knowledge of human society, the relationships between human beings and nature and society, social phenomena and the development of human society. Political, ethical, military, legal, and other social scientific ideas gradually took shape during this phase. In the second phase of qualitative research (since the 1960s), qualitative research has largely shaped modern social science, which includes observing, recording, comparing, classifying, and summarizing relationships between facts, and verifying the results. This research paradigm is continuously being developed.

8.3 The Second Research Paradigm

The second research paradigm arguably fast-tracked the development of modern science, where a number of hypothetical conditions are generalized in scientific 'laws'. For example, Newton constructed the three laws of motion based on three major assumptions: (i) spatial assumption (space is void and flat rather than curved); (ii) the mass point assumption (matter has only mass and no natural properties such as electromagnetism, geometry, etc.); and (iii) inertia assumption (matter is inherently partly in motion and partly at rest). The most distinctive feature of the second research paradigm is the use of models to simplify research problems and to draw conclusions through careful logical reasoning and calculations.

The second research paradigm for social science is quantitative research. Influenced by the natural sciences, more and more social science researchers are favouring mathematical methods, mainly because mathematics, as a precise and unambiguous language, can further extend the researcher's deductive reasoning and provide more convincing results and findings. Quantitative methods are often presented in the forms of hypothesis development and validation.

8.4 The Third Research Paradigm

The human brain is limited in its computational capacity, and when the amount of computation involved in a research problem becomes too large, the human brain might be overwhelmed. Against this background, computers emerged. In 1946, the first electronic computer 'ENIAC (Electronic Numerical Integrator and Computer)' was developed and introduced, making it possible to simulate complex scientific experiments. The power of computers allows people to simulate complex phenomena, and deduce and explain more complex phenomena, such as nuclear tests simulation and weather forecasting. As computer simulations increasingly replace experiments, computational science is becoming the norm in scientific research, a third research paradigm.

Computer simulations also open up ideas for social science research. Based on mathematical methods, computer technology, statistical science, information science, and systems theory, social science researchers use computer programming to simulate in a virtual environment social phenomena, the state of development and future trends of social changes that may occur in the real world. The application of computer simulation in the social sciences removes the limitations of social science research objects that cannot be subject to experiment or repeated, enables real-world research and implementation environments that are generally costly or impossible to obtain, and provides effective methods and tools for understanding and grasping the structure and function of complex social systems (Mi et al. 2018).

Meanwhile, researchers are becoming aware of the interdisciplinary nature of social science. Accordingly, mixed methods research emerges. Mixed methods research is based on pragmatism at the philosophical level, and concentrates on collecting, analysing, and mixing both quantitative and qualitative data for the same research aim at the methodological level (Creswell and Clark 2017). Mixed methods research is becoming increasingly important in several scientific areas, including construction management (Abowitz and Toole 2010; Zou et al. 2014), behavioural sciences (Lopez-Fernandez and Molina-Azorin 2011), information system Venkatesh et al. 2013, and organization management (Myers and Oetzel 2003). Social science computing and mixed methods research together constitute the third research paradigm of social science.

8.5   The Fourth Research Paradigm

With the explosive growth of data volumes and the rapid increase in the complexity of research problems, research is entering to the fourth research paradigm of 'data-intensive scientific discovery'. Humanity is accumulating previously unimaginable data in digital form that will help bring profound changes in scientific research and insight (Hey et al 2009). Different from the third research paradigm, the fourth paradigm does not emphasize the need to start with research hypotheses and theoretical underpinnings, but rather focuses on how to discover new knowledge and theories from the vast amount of data. For instance, biologists use computers to collect and analyse large amounts of data such as DNA, proteins, and bioinformatics to reveal the potential modes of action and mechanisms of occurrence of different diseases at the molecular level; astrophysicists classify galaxies by analysing vast amounts of astronomical data.

'Data-intensive scientific discovery' has also had a huge impact on the social sciences. The rapid development of the Internet and metaverse, social networks, the Internet of Things, call records, and radio frequency identification (RFID) have broadened the outreach of social science research. These methods of data collection are more efficient than traditional questionnaire surveys and interviews and also minimize social bias in data collection. Driven by big data, the object of social science research has shifted from traditional human involvement in social systems and processes to a network of data formed by the interactions of parallel systems in the real and virtual worlds (Mi et al. 2018).

8.6   Evolution of Research Paradigms

There is criticism about each of the research paradigms, which is the driving force for the evolution of research paradigms. Research paradigms have been continuously evolving but they are not independent of each other. Instead, there is an evolutionary relationship between them, as shown in Figure 8.1.

In natural science, empirical evidence (the first research paradigm) is limited by experimental conditions and low accuracy. To address the shortcomings of the first research paradigm, researchers attempted to propose a simplified experimental model (establishing ideal experimental conditions) and to obtain inductive summaries through meticulous logical deduction and mathematical calculations. At this point, the second research paradigm emerged. As research develops further, the computational demands of research become increasingly demanding, and the limited computational power of the human brain requires the help of computers. The advent of computational science (the third research paradigm) has allowed researchers to simulate and extrapolate complex phenomena at low cost. However, with the rapid increase in the complexity of research problems and the explosive growth of data volumes, the third research paradigm seems less capable than desirable. Thus, data-intensive scientific discovery (the fourth research paradigm) emerged.

In social science, qualitative research (the first research paradigm) is usually conducted in the early stages of study when researchers do not have a systematic understanding of that being researched. With discussion, open-ended answers and explanations shared by participants researchers could gain information and ideas beyond what they could get from survey and secondary data (AlModhayan 2016). Despite these advantages, there

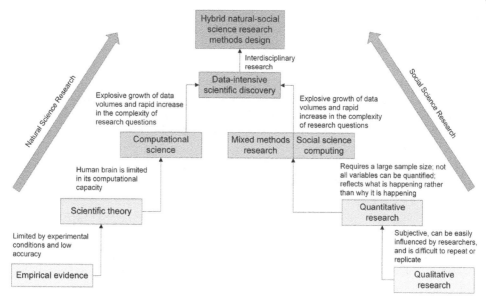

Figure 8.1 Evolution of research paradigms.

is common and constant criticism of qualitative research. Many researchers argue that qualitative study is subjective and can be easily influenced by researchers. In reality, qualitative research is difficult to repeat or replicate. Even with the same procedure and sample, different researchers may have different findings due to different perceptions. Furthermore, critics also pointed out that limited samples in qualitative research cannot represent populations; this is seen as another weakness.

Quantitative research (the second research paradigm) can not only produce exact numerical results but also present uncertainty and fuzziness, e.g., Monte Carlo simulation and fuzzy mathematics. Although the quantitative research method is genuinely popular with researchers, it is not free from criticism. Carr (1994) stated that most quantitative research does not explore phenomenon in a natural setting because it aims to control or eliminate the impact of extraneous on the result of research. Furthermore, quantitative research usually requires a large sample size to ensure the accuracy and representativeness of the results, which is expensive and time consuming (Morgan and Smircich 1980). However, due to resource constraints, the large-scale quantitative research sometimes becomes impossible. In addition, it is difficult for researchers to quantify some variables (e.g., emotion and culture) under the existing level of knowledge. Lastly, while quantitative research produces exact numerical results, it reflects what is happening rather than why it is happening (Zou et al. 2014) and, therefore, needs to be interpreted further by the researchers.

To achieve complementary strengths in qualitative and quantitative research, mixed methods research was developed as a kind of research design that integrates qualitative and quantitative methodologies in the same research. Many researchers believed that the use of mixed methods research can help enrich and improve the understanding of various phenomena of interest that cannot be fully understood by a single methodology (Lopez-Fernandez and Molina-Azorin 2011; Venkatesh et al. 2013). Meanwhile, social science computing provides new ideas for the study of the complex structure and function of social systems. Similar to the natural sciences, the first three research paradigms in the social sciences were unable to meet the needs of social science research as the volume of data grew by leaps and bounds, so the social sciences moved into a fourth paradigm, i.e. data-intensive scientific discovery.

As shown in Figure 8.1, with continuous evolution and development and application of ICT, AI, and data science, together with focuses on being problem-oriented, interdisciplinary-combined and data-driven, research is entering the fifth research paradigm of hybrid natural-social science methods research design, as discussed in the next section.

8.7 The Fifth Research Paradigm

The fifth research paradigm combines natural science and social science (Figure 8.2). The use of the fifth paradigm is based on four principles.

1) Problem-orientation. The research should commence from a clearly identifyied research problem. Researchers need to consider whether the research problem is complex enough to involve both the natural and social sciences methods. Researchers should also discuss the research problem in terms of complexity, systematicness, and revolutionization. In this context, complexity refers to situations where some systems display behavioural phenomena that are completely inexplicable by any conventional analysis; systematicness means a hierarchical whole with elements of different dimensions at

Figure 8.2 Characteristics of the fifth research paradigm.

different levels forming a certain order and a clear logical relationship between elements at the same level and between different levels; and revolutionization means having a disruptive impact on existing theories and methods.

2) Philosophical standpoint. This may be specified from the perspective of ontology, epistemology, and axiology (as discussed in Chapter 1). It is important for researchers to clarify the philosophical standpoint, so that appropriate research methodology can be selected.

3) Interdisciplinarity. Researchers should clarify that the research involves multiple disciplines, which include social science and natural science disciplines. Researchers need to focus on the intersection of which disciplines are involved, how the disciplines intersect, and the criteria, processes, and multiple scopes, or diversity of scope, for this interdisciplinarity. Criteria refer to the standard on which the interdisciplinarity should be based. Processes refer to a program in execution for interdisciplinarity. Multiple scopes mean the diversity of the research coverage.

4) Data-driven. This extends of the fourth research paradigm. Researchers need to explore in-depth how data and research methods from different disciplines can be hybridized to address the research problem. Algorithms, computing power, and multisource data are at the heart of the fourth principle. Algorithms are a set of well-defined process or rules to be followed in calculation. Computing power refers to the computing capacity of a device. Multisource data refers to data that are diverse in terms of sources and types.

8.8 Application of the Fifth Research Paradigm

Modes of integration of natural science and social science in the fifth paradigm include but are not limited to using the laws of natural science to explain the social phenomena; using data collection technologies from the natural science to broaden the volume, velocity, variety, and value of data analysis in the social sciences; and using social research methods to test the effectiveness and measure the benefits of natural research outcomes.

In the first mode, researchers could use principles or formulas from the natural sciences to explain social phenomena. For example, in a social team setting research, to answer the question of how to achieve maximum team synergy and generate maximum outcomes, we could use Newtonian laws to model the contribution of each member in the team as an individual force to explain and quantitatively calculate the resultant force, which would be similar to modelling and calculating the resultant forces of a number of forces. By moving the positions and directions of each force, eventually moving all forces into the same direction and position, so achieving the maximum resultant force (i.e., the maximum team synergy). Another example would be to use mathematical laws, formulas, equations, and differentiation to describe and model social phenomena, which then can be calculated or simulated. Research into human–cyber–physical systems falls within this paradigm.

In the second mode, researchers could use advanced data collection techniques and tools in the natural sciences to obtain objective social research data. For example, electroencephalogram (EEG) data can be collected by using EEG technology to assist in the analysis of workers' fatigue, hazard recognition, and work discipline during operations. Compared to data collected through questionnaires, data collected by EEG is a more objective in response to the cognitive state of the respondents. More information on applying EEG technology is in Chapter 6.

In the third mode, researchers could investigate the application of new natural science research outcomes from social science perspectives (e.g., policy, economic, and social perspectives). For instance, the application of photovoltaic power generation technology could be explored in terms of economic benefits, carbon reduction benefits, government subsidies, and technology acceptance. At the same time, market and economic developments could also promote technological innovation in photovoltaic power generation.

8.9 Summary

The time for integrating natural science and social science research methods has come. This integration forms the fifth research paradigm. Some research has already used different research methods drawn from natural science and social science to solve complex research problems. This is a natural progress in research methodology domain. However, in the process of evolution of research paradigms, this paradigm by no means replaces the earlier ones, instead, the different research paradigms continue to evolve and improve. The selection decision of research paradigm(s) depends on the nature of research problem(s) and question(s). From a pragmatic perspective, no matter what methods are used, as long as these methods are effective in resolving the research problems and helping achieve research aims, they are suitable. In summary, the fifth research paradigm is a natural extension and progression of the fourth research paradigm. We believe that the fifth research paradigm will gain more and more attention and application, and readers are encouraged to consider this hybrid methods research design and implementation.

Review Questions and Exercises

1 What is your understanding of the five research paradigms?
2 What drives the evolution of research paradigms?
3 What is computational experimental research?
4 What are the roles and functions of data science in the evolution of research paradigms?
5 Under what situation would it be suitable to use hybrid natural-social science methods?

References

Abowitz, D.A. and Toole, T.M. (2010). Mixed method research: fundamental issues of design, validity, and reliability in construction research. *Journal of Construction Engineering and Management* 136 (1): 108–116.

AlModhayan, M. (2016). Merits and demerits of qualitative research. Available from: https://www.linkedin.com/pulse/merits-demerits-qualitative-research-mohammed-almodhayan; accessed 8 April 2023.

Carr, L.T. (1994). The strengths and weaknesses of quantitative and qualitative research: what method for nursing? *Journal of Advanced Nursing* 20 (4): 716–721.

Creswell, J.W. and Clark, V.L.P. (2017). *Designing and Conducting Mixed Methods Research.* SAGE Publications.

Hey, T., Tansley, S., Tolle, K.M., and Gray, J. (2009). *The Fourth Paradigm: Data-intensive Scientific Discovery.* Redmond, WA: Microsoft Research.

Lopez-Fernandez, O. and Molina-Azorin, J.F. (2011). The use of mixed methods research in the field of behavioural sciences. *Quality & Quantity* 45 (6): 1459–1472.

Mi, J.N., Zhang, C.P., Li, D.Y., and Lin, T. (2018). The fourth research paradigm: transforming social science research driven by big data. *Xuehai* (2): 11–27. in Chinese.

Morgan, G. and Smircich, L. (1980). The case for qualitative research. *Academy of Management Review* 5 (4): 491–500.

Myers, K.K. and Oetzel, J.G. (2003). Exploring the dimensions of organizational assimilation: creating and validating a measure. *Communication Quarterly* 51 (4): 438–457.

Venkatesh, V., Brown, S.A., and Bala, H. (2013). Bridging the qualitative-quantitative divide: guidelines for conducting mixed methods research in information systems. *MIS Quarterly* 37 (1): 21–54.

Zou, P.X.W., Sunindijo, R.Y., and Dainty, A.R.J. (2014). A mixed methods research design for bridging the gap between research and practice in construction safety. *Safety Science* 70: 316–326.

9

Journal Article Writing and Publishing

9.1 Introduction

Writing articles for publication in journals and at conferences is a process that all researchers go through, as part of the job or career performance and outcomes. It is important to discuss publication in an open and transparent manner. Publishing an article demonstrates someone's research and innovation ability and allows them to share their hard work. Publishing articles in international journals not only allows researchers to exchange research outcomes with more peers but also to achieve international influence. From the point of view of personal improvement, this process can help improve problem solving and logical thinking abilities. In this chapter, we discuss the structure of a journal article, the writing and submission processes, and some key points on responding to reviewers' comments.

9.2 Structure of a Journal Article

There are several types of articles that may be published in a journal, such as original research paper, review paper, book review, letter, technical note, and forum. An original research paper usually consists of several sections, including title, abstract, keywords, introduction, literature review, research methods, results and discussion, conclusion, acknowledgments, and references. This may vary across academic disciplines. One example of an original research paper is shown in Example 9.1.

Example 9.1   Discovery of new safety knowledge from mining large injury dataset in construction projects

1. Introduction
2. Literature review
   2.1 Overview of construction injury research
   2.2 Construction safety research methods

Research Methodology and Strategy: Theory and Practice, First Edition. Patrick X.W. Zou and Xiaoxiao Xu.
© 2023 John Wiley & Sons Ltd. Published 2023 by John Wiley & Sons Ltd.

2.3 Summary and point of departure
3. Research methods and processes
   3.1 Description of the dataset
   3.2 Data preparation
   3.3 Data analysis methods and processes
      3.3.1 Statistical analysis
      3.3.2 Data visualization
      3.3.3 Association rule mining
4. Results and discussion
   4.1 Time of Injury
      4.1.1 Injury time of the year
      4.1.2 Injury time of the day
   4.2 Bodily locations, organs, and systems of injuries
      4.2.1 Bodily locations and organs
      4.2.2 Injured body systems
   4.3 Age distribution of the injured workers
   4.4 Causes and nature of injury relating to body parts
      4.4.1 Causes of injury relating to body parts
      4.4.2 Nature of injury relating to body parts
      4.4.3 Results of association rule mining (ARM) analysis
   4.5 A decade long changing trends of incidents
5. Strategies for improving safety management
   5.1 Providing personalized safety training
   5.2 Conducting short safety meetings at strategic points
   5.3 Implementing tailored first-aid training and medical preparation
   5.4 Making wearing personal protective equipment mandatory on site
   5.5 Enhancing real-time on-site monitoring by applying emerging digital technologies
6. Research implications and contributions
7. Summary, future research directions, and conclusion
Acknowledgement
References

This example is an original research paper published in the journal of *Safety Science* by Xu and Zou (2021). The structure of this paper can be summarized as: 'Pursuant to the introductory section is a literature review (Section 2). A detailed description of the research methods, including description of the dataset, data preparation, and data analysis methods and processes, is presented in Section 3. Section 4 shows the results of the mining of the injury database. Section 5 proposes five strategies for improving construction safety performance, followed by research implications and contributions (Section 6). Section 7 concludes the study, with main findings and future research directions'. For the full paper, please see 'Discovery of new safety knowledge from mining large injury dataset in construction projects' published in *Safety Science* (https://doi.org/10.1016/j.ssci.2021.105481).

The typical structure of a review-based research paper could include the following components: title, introduction, background information, literature selection, evaluation of literature, informatic analysis, discussion, conclusion, and references. The discussion section should include current body of theories, current body of methods, theoretical and practical implications, current research limitations, and future research directions. However, there is no fixed structure for review-based research papers. Apart from the first section (introduction) and last section (conclusion), researchers can design the structure by themselves. In any case, they should not be divorced from the basic logic of review papers: giving a picture of the current state of research in the field, finding the current research gap, and proposing future research directions. Two examples of review papers are shown in Example 9.2 and Example 9.3.

Example 9.2   A mixed methods design for building occupants' energy behaviour research

1. Introduction
2. Analysis of occupant energy behaviour research methods
    2.1 Article selection
    2.2 Research methods adopted in occupant energy behaviour research
    2.3 Discussion
3. Overview of research methodologies
    3.1 Philosophical understanding of research methodologies
    3.2 Quantitative research
    3.3 Qualitative research
    3.4 Mixed methods research
4. Mixed methods research design for occupant energy behaviour research
5. Conclusion
Acknowledgement
References

Example 9.2 is a review paper published in the journal of *Energy and Buildings* by Zou et al. (2018a). This review paper aimed to design a framework of mixed methods research in the field of energy-related occupant behaviour on the basis of the existing framework of mixed methods research in other relevant areas. In the Introduction, this paper listed three objectives: (i) to understand the state of the art of research methods used in building occupants' energy behaviour research; (ii) to identify conditions for application of mixed methods research design in building occupants' energy behaviour research; and (iii) to develop a framework and process for conducting mixed methods in building occupants' energy behaviour research. Section 2, Section 3, and Section 4 correspond to the three objectives mentioned in the introduction. For the full paper, please see 'A mixed methods design for building occupants' energy behaviour research' (https://doi.org/10.1016/j.enbuild.2018.01.068).

Example 9.3   Review of 10 years research on building energy performance gap: Life-cycle and stakeholder perspectives

1. Introduction
2. Research methods
    2.1 Retrieving publications
    2.2 Analysing contents using NVivo*
3. Summary of previous building energy performance gap research
    3.1 Building types
    3.2 Strategy spectrum
    3.3 Building life-cycle
    3.4 Energy-related stakeholders
    3.5 Influencing factor
4. Root causes of building energy performance gap
    4.1 Design and simulation related causes
    4.2 Construction related causes
    4.3 Operation related causes
5. Strategies for closing building energy performance gap
    5.1 Design concepts
        5.1.1 Passive and active design
        5.1.2 Human-in-the-loop
    5.2 Technology and method
        5.2.1 Technology and method for calculating energy consumption
        5.2.2 Technology and method for data collection and analysis
        5.2.3 Technology and method for occupant behaviour modelling and simulation
        5.2.4 Technology and method for building mechanical system controlling
    5.3 "Soft" measures
        5.3.1 Policy
        5.3.2 Rating system, benchmarking, and standards
        5.3.3 Collaboration and communication
    5.4 Limitations of current strategies
6. Future research needs
7. Conclusion

Example 9.3 is a review paper also published in *Energy and Buildings* (Zou et al. 2018b). The second section described the research method used in this research. The third section provided a panorama of building energy performance gap research. Subsequently, Section 4 and Section 5 discussed the root causes and corresponding strategies, respectively. After analysing root causes and their corresponding strategies, the research gap was presented. Section 6 discussed future research directions. For the full paper, please see 'Review of 10 years research on building energy performance gap: Life-cycle and stakeholder perspectives' (https://doi.org/10.1016/j.enbuild.2018.08.040).

For review papers, we have three key recommendations:

1) The paper should include recent articles.
2) Articles collected should be sufficient to cover the contents of the review.
3) Researchers should provide new insight or knowledge generated rather than common sense from the research.

9.3 Writing a Journal Manuscript

We use the term manuscript instead of paper or article here because, normally, a piece of writing for submission to a journal is called manuscript and once accepted by the journal it becomes an article of the journal. And the term paper is used often in conference proceedings. However, there is no strict rule about the use of these terms.

9.3.1 The Title

A good title can help the paper attract more readers. We have seven points to help researchers write a title:

1) Reflect the research aim.
2) Make sure it is attractive.
3) Clearly indicate the contents and focus of the research.
4) Make it is clear and concise (usually no more than 25 words).
5) Make it descriptive and include keywords.
6) Refine the title at the end of research.

9.3.2 The Abstract

An abstract is usually located at the top of a research paper. The main purpose of the abstract is to give a clear account of the background and rationale for the choice of topic, the ideas and values of the paper, so that readers can obtain the important information in the article in time without reading the full text. The abstract consists of the basic elements of the background of the research, the purpose of the research, the research methodology, and the main conclusions drawn from the study. The contents of an abstract are derived from the rest of the manuscript. Thus, once researchers have finished other parts of the manuscript, it is easy to write the abstract. The following is an analysis of an abstract (Example 9.4) from Zou et al. (2018a) 'A mixed methods design for building occupants' energy behaviour research' (https://doi.org/10.1016/j.enbuild.2018.01.068).

Example 9.4

Abstract: Occupant behaviour is viewed as a main source causing the building energy performance gap between the predicted and the actual consumption **[Sentence 1]**. The nature of occupants' energy behaviour research requires a combination of social science and natural science, which indicates that a mixed methods design would be useful

---

(Continued)

(Continued)

[Sentence 2]. However, researchers often do not know when and how mixed methods approach should be used [Sentence 3]. To fill this gap, this paper first reviewed the research methods adopted in 230 relevant articles published in the past decade [Sentence 4]. The results show that 83.48% of articles applied quantitative methods, followed by mixed methods (5.22%) and qualitative methods (0.87%) with rest being pure review or conceptual papers [Sentence 5]. This shows that researchers in the field of occupant behaviour mainly adopt the objectivist philosophical position [Sentence 6]. Subsequently, an in-depth analysis of research methodologies was conducted in relation to worldviews and philosophical assumptions, and the advantages and disadvantages of qualitative and quantitative methods were discussed [Sentence 7]. Finally, a mixed methods research design framework was proposed as a point of departure for researchers to gain a comprehensive understanding of mixed methods design [Sentence 8]. It is expected that this framework could help researchers develop a proper mixed methods research design according to the nature of their research problem [Sentence 9].

In **Sentence 1**, the authors present the research background. It highlights the importance of occupant behaviour in research as it is the main reason of building energy performance gap. Subsequently **Sentence 2** and **Sentence 3** provide the research needs and research gap. To fill the identified research gap, **Sentences 4–7** present the research methods, research process, and research findings. In **Sentence 8**, the authors indicate the achievement of the research. In **Sentence 9**, the authors present the contribution of the research.

9.3.3 The Keywords

A maximum of five keywords is usually required for journal papers. General and plural terms and multiple concepts should be avoided. The keywords should reflect the main theme, methods, or findings of the research. The keywords should also allow the maximum degree of 'searchability', i.e., allow the article to be widely searched on most search engines.

9.3.4 The Introduction

The introduction is the most important section in the journal paper and it may be the most difficult part to write. The introduction needs to tell readers (i) the research background, (i) the importance of this research, (iii) the research problems, and (iv) research aim and objectives.

The normal practice is to start with research background in the introduction section. If the research field is specific and most of the readers have a high level of background knowledge, researchers can start with specific information. If the paper is likely of interest to a wider range of readers, researchers should start with more general background information (Hilary 2010). Example 9.5 is an example of a 'research background' cited from Zou et al. (2017) 'Cloud-based safety information and communication system in infrastructure construction' (https://doi.org/10.1016/j.ssci.2017.05.006).

Example 9.5

Working in the construction industry is considered more risky than in many other industries [**Sentence 1**]. Statistical figures have suggested that both the rates of serious injury compensation claims and fatalities in construction have been consistently higher than their corresponding overall industrial average rate (Zou and Sunindijo, 2015) [**Sentence 2**]. For example, according to Safe Work Australia's industry statistics reports (Safe Work, 2014a, Safe Work, 2014b), the Australian construction industry recorded 402 fatalities and 140,448 serious injury claims from 2003 to 2013 [**Sentence 3**]. These records accounted for 14% and 10% respectively of the total national work-related fatalities and serious injury compensation claims, which were the third highest amongst all industries over the 10-year period [**Sentence 4**]. The need for better safety performance in the construction industry is eminent [**Sentence 5**]. Recent studies have reported that information technology (IT) innovations can be implemented in construction safety management through various applications including safety in design (Eastman et al., 2009, Hayne et al., 2014, Hu et al., 2008, Zhang and Hu, 2011), safety site planning (Azhar et al., 2012, Wang et al., 2014, Bansal, 2011, Naik et al., 2011, Cheng and Teizer, 2013, Guo et al., 2013, Huang et al., 2007, Lai and Kang, 2009, Li et al., 2003), safety risk management (Hartmann et al., 2012, Melzner et al., 2013, Zhang et al., 2015, Zhang et al., 2013, Bansal and Pal, 2009, Baertlein et al., 2000, Talbot and Nichols, 1999, Hadikusumo and Rowlinson, 2004, Hadikusumo and Rowlinson, 2002, Park and Kim, 2013), safety knowledge database (Park and Kim, 2013, Vossebeld and Hartmann, 2014), safety monitoring (Kimmance et al., 1999, Lee et al., 2009, Teizer et al., 2010, Yang et al., 2012, Han et al., 2009) and safety training (Park and Kim, 2013, Guo et al., 2012) [**Sentence 6**]. A review conducted by Skibniewski (Skibniewski, 2014) on IT innovations in construction safety management from 2006 to 2014 reported that sensors and sensor-based systems for safety, robotics and manipulators for safety and information analysis and management technologies are the major topics among the studies [**Sentence 7**]. The Skibniewski (2014) review also showed that IT related topics consisting of building information modelling (BIM) for safety in design and safety risk management (Eastman et al., 2009, Hayne et al., 2014, Hu et al., 2008, Zhang and Hu, 2011, Azhar et al., 2012, Wang et al., 2014, Hartmann et al., 2012, Melzner et al., 2013, Zhang et al., 2015, Zhang et al., 2013, Park and Kim, 2013, Vossebeld and Hartmann, 2014), sensor-based location tracking systems such as the Global Positioning System (GPS) for safety monitoring (Cheng and Teizer, 2013, Baertlein et al., 2000, Park and Kim, 2013, Lee et al., 2009, Teizer et al., 2010, Yang et al., 2012) and computer-aided design (CAD) were proven to be useful for creating better safety information management system [**Sentence 8**]. Other studies reported that geographic information system (GIS) is also a promising IT innovation for improving construction safety site planning and risk management (Wang et al., 2014, Bansal, 2011, Naik et al., 2011, Bansal and Pal, 2009, Kimmance et al., 1999) [**Sentence 9**].

In **Sentence 1** and **Sentence 2**, the authors directly show a grim situation: safety accidents in the construction industry are more common than other industries. In **Sentences 3** and **4**, the authors used data and research references to support the background and the research significance. **Sentence 5** is a turning point and leads to strategies for avoiding safety accidents. Subsequently, **Sentences 5**, **6**, **7**, **8**, and **9** provided a brief description of current technology for construction safety management.

The research problem is the core of a research paper and needs to be clearly stated. The most common way is to identify a research problem from previous research papers' future research directions and suggestions or limitations. Research problems may also be derived from conversation with practitioners, and verified by literature. The following is the analysis of another 'research problem' (Example 9.6) cited from Zou et al. (2017) 'Cloud-based safety information and communication system in infrastructure construction' (https://doi.org/10.1016/j.ssci.2017.05.006).

Example 9.6

There are several drawbacks of current commercial systems: First it is hosted by a third party, which means a fee need to be paid and limited space may be provided; Second it is time-consuming and uneasy to customise the system to match with your organisational and project settings; Third, it is difficult to provide real-time communication; Forth, it is uneasy to use an App on mobile devices; Fifth, most of these systems are general systems and not specific to safety management, which means some safety functions may not be included; Sixth, many of these systems do not have GIS and GPS function built in, which make them unsuitable for horizontal type infrastructure projects such as road and rail construction projects [**Sentence 1**]. As such, there is a need to undertake research and develop a cloud-based safety information and communication system for infrastructure construction [**Sentence 2**].

Many researchers recognised that safety information flow is the core of an effective safety management system (Kirwan, 1998, Bottomley, 1999, Cheng et al., 2012, Fernández-Muñiz et al., 2009, Wilson and Koehn, 2000, Zhou et al., 2015), though there are only few reports that have addressed the need for real-time communication of safety information between construction workers (El-Saboni et al., 2009, Nuntasunti and Bernold, 2006) [**Sentence 3**]. An effective system does not only have positive impact on safety of construction workers but also ensure that employers comply with relevant legal requirements (Fernández-Muñiz et al., 2007) [**Sentence 4**]. In contrast to the advancing of IT applications in construction safety as described above, current safety information and communication management systems are not fully utilising currently available IT [**Sentence 5**]. The current system is still by and large utilising paper-based documents to capture and store safety information [**Sentence 6**]. The paper-based documents are then scanned and stored in the relevant safety directory [**Sentence 7**]. There are three issues of the system: First, safety information is not available in real-time; Second, it takes considerable administrative efforts to digitise paper-based documents and to file them correctly in the safety directory; Third, it is difficult to track the digitised information in large-scale infrastructure construction projects [**Sentence 8**]. Hence, there remains a need for more efficient and effective safety information management systems for construction [**Sentence 9**].

Sentence 1 lists the drawbacks of current commercial systems while **Sentence 2** points out the research needs. Based on the research needs, **Sentences 3**, **4**, **5**, **6**, and **7** discuss the current studies on the safety management system. In **Sentence 8**, the authors identify the limitations of current researches. In **Sentence 9**, the authors propose the research problem. In a journal paper, research problem requires references to properly acknowledge the sources of ideas.

Once the research problem is identified, researchers can provide a transition between the research problem and research aim. In general, the research problem and the essence of the research aim are the same. Example 9.7 shows the relationship between research problem and research aim cited from Xu and Zou (2020) 'Analysis of factors and their hierarchical relationships influencing building energy performance using interpretive structural modelling (ISM) approach' (https://doi.org/10.1016/j.jclepro.2020.122650).

Example 9.7

Question: What are the interrelationships among the influencing factors affecting building energy performance gap?

Aim: the research aims to identify the key factors and explore the interrelationships among the factors affecting building energy performance gap.

Subquestions:

- What are the factors affecting building energy performance gap?
- What are the interrelationships among these factors?
- What are the driving and dependence power of each factor?
- How are the factors and their interrelationships addressed?

Objectives:

- Identify representative factors affecting building energy performance gap.
- Determine and evaluate the interrelationships among these factors.
- Identify the driving and dependence power of each factor.
- Propose strategies for addressing the factors and their interrelationships.

Researchers can use the format shown in Example 9.8 to present research aim and objectives.

Example 9.8

Therefore, this research aims to [.......]. Based on this aim, three research objectives are formulated as follows:

1. [Objective 1]
2. [Objective 2]
3. [Objective 3]

Researchers can also mention the structure of their paper at the end of the introduction section, as shown in Example 9.9.

Example 9.9

The remainder of this paper is structured into four sections. Pursuant to this introductory section is [.......]. [......] is utilised to explore [...]. Section 3 details the process of [......]. Section 4 conducts [......]. Conclusion and future research are given in section 5.

9.3.5 The Literature Review

A literature review is a systematic, comprehensive, and critical overview of a research topic area over a certain period of time (Booth et al. 2012). Normally, there is a particular need or purpose for undertaking a literature review, for example to comprehend and synthesize the theory and knowledge domains and progress or to establish the needs, significance, and aims of new research, particularly as in the context of this chapter (journal article writing). A good literature review should be retrospective, evaluative, instructive, comprehensive, and critical. It should describe the current state of development and trends in the field of research, point out the breakthroughs, and draw out future research directions.

A literature review starts with searching for and collecting sufficient literature. The search for critical and foundational literature should be based on three keywords: authority, relevance, and applicability of the literature. How to find the relevant literature is particularly important. For literature, search keywords can be determined according to research context, research subject, research object, and research perspective. For instance, if you were given a topic entitled 'Factors influencing the energy use behaviour of public building', the first step is to define the framework of the topic: the research context is building energy efficiency, the research subject is building users, the research object is energy use behaviour, and the research perspective is behavioural influencing factors. The second step is to decide the keywords. The keywords in this case would be public buildings, users' energy use behaviour, and influencing factors. The third step is to rank the keywords from largest to smallest according to the scope. The final step is to find relevant literature on the topic for each keyword. In this way we can broaden our search of the relevant literature within library and internet databases.

The good way to identify critical and foundational literature is to look at the division of the journal titles. The most commonly used journal partition is the JCR partition. Journals ranked in the top 25% are in Zone 1, journals ranked between 25% and 50% are in Zone 2, journals ranked between 50% and 75% are in Zone 3, and journals ranked between 75% and 100% are in Zone 4. Another method is to look at the number of citations of the journal articles. A higher number of citations means that the article is more applied and inspiring, which means that it could be more classic and valuable to read. It is advised to use software to manage your literature.

The next step is to read and analyse the literature (the many journal articles and conference papers) relevant to the research topic. Siegel (2020) suggested five steps for reading and analysing literature.

1) Evaluate the article, that is, read the title quickly and find the keywords.
2) Read the abstract quickly, look for the keywords, and understand the abstract; then, carefully read the introduction. The introduction is often a relatively easy-to-understand

part of the article and can be used to integrate the background information. The introduction will generally contain many references.

3) Understand the way the article deals with the problem. Carefully read and understand the lines and tables, which may take several times to understand, but the information given is enormous.

4) Read the full text for the first time, skim the abstract and introduction again, and skim the methods section unless they are involved in the research problem. Read the results, discussion sections, as well as research charts and tables.

5) Reread the full text and make notes, including any words you do not understand, or any doubts or irrationalities. Refer to the reference materials for the knowledge points you do not understand, supplement the knowledge points in time, and read the abstract again before finishing reading the article.

It is suggested to make notes and compile the information in the literature as a summary list in chronological order. From the summary list, researchers can identify common ideas and linkages between the publications. Table 9.1 shows how a table might help researchers summarize the relevant publications.

There is a wide range of software available to assist researchers in conducting literature review, as listed in alphabetic order in Table 9.2.

Table 9.1 Summary list of publications.

No.	Authors and year	Problem statement/ research question(s)	Material and methods	Results/Findings	Limitations
1.	Xiaoxiao Xu and Patrick X.W. Zou (2021)	It is difficult to extract patterns and gain new knowledge for improving safety performance.	A large set of injury and incident data recorded by construction companies over a period of 11 years was used in this research. Statistical analysis, visualization analysis and association rule mining (ARM) methods were applied to mine and analyse this large dataset.	"Injury time", "injured body locations, organs and systems", "age distribution of injured", "causes, nature and relationships of injury, and injured body parts", "changing trends of incidents" were analysed to discover new safety knowledge.	Although this study has applied statistical analysis, data visualization, and association rule mining methods to discover injury-related knowledge, there are other data mining methods, for example, cluster detection, decision tree, and classification could also be used to discover more knowledge. Future research could also focus on how to remove the root causes of injury that have been identified in this research.

Table 9.2 Software for analysing literature.

Software	Descriptions	Websites
Bibexcel	Bibexcel is a professional and easy-to-use bibliometric analysis software. It provides users with bibliometric analysis, citation analysis and coword analysis, which can be used as a tool for researchers in their daily research and for science and technology management and policy analysis.	http://homepage.univie.ac.at/juan.gorraiz/bibexcel/bibexcel.exe
CiteSpace	CiteSpace is a visual literature analysis software that uses the interrelationship of the literature to track hotspots and trends in a field of study in the form of a scientific knowledge map, understanding the research frontiers and key paths of evolution, important literature, authors and institutions. Its "emergence" helps researchers to keep up with the hot spots in the field.	https://citespace.podia.com
Histcite	Histcite is a bibliographic analysis software used to process bibliographic information exported from the web of science. It can help researchers quickly grasp the historical development of the literature in a field and identify highly cited key articles and highly cited key authors. It also makes it easy to map the historical relationships of the literature in the field, making it easy to see the development, relationship, and researchers in the field at a glance.	https://histcite.software.informer.com
VOSviewer	VOSviewer is a tool for visualization and analysis of literature, i.e., for mapping scientific knowledge through the construction and visualization of relationships in "network data" (mainly knowledge units of literature), showing the structure, evolution, cooperation, and other relationships in the knowledge domain, suitable for large-scale sample data.	www.vosviewer.com

In addition to the software listed in Table 9.2, NVivo®, as a qualitative data analysis computer software, is a useful tool to analyse the literature (Li et al. 2014; Lu and Yuan 2011). All publications imported into NVivo® can be treated as 'Sources' and analysed using the 'Node' function. A node is a collection of references (including sentence, paragraph, the whole paper) about a specific topic, theme, or relationship. The references about the same theme can be gathered to a node by 'coding'. Take a paper titled 'A virtual reality integrated design approach to improving occupancy information integrity for closing the building energy performance gap' (Niu et al. 2016) published in the journal of *Sustainable Cities and Society* as an example. The sentence 'In the operation stage, occupants may not perform in accordance with building designers' design assumptions' is related to operation, occupant, and design assumption, thus three nodes are created, namely 'operation', 'occupant', and 'design', and the sentences under them can be coded. In most cases, nodes may have a structure of two or more levels. About the occupant, researchers could create a three-level node structure where the first level includes 'Operation', the second level includes

'Occupant', and the third level contains 'Occupant characteristics', 'Occupant comfort', 'Occupant experience', and 'Occupant behaviour'. Then researchers analyse and code the contents under 'Occupant characteristics', 'Occupant comfort', 'Occupant experience' or 'Occupant behaviour'. All sources can be coded using this approach. Initial codes might be iteratively revised and refined throughout the coding process (Li et al. 2014). In order to ensure the reliability and validity of the analytical results, several rounds of coding need to be conducted.

Regarding the structure and presentation of a literature review, some people use a single sentence pattern, such as 'XXX pointed out...', 'XXX found that...', 'XXX hold the view that...'. This kind of sentence pattern will make the whole literature review look repetitive, diffuse, and incomplete. Readers will probably feel bored. Instead, researchers should use different sentence patterns. In fact, there are several ways to structure a literature review: structure according to chronology, structure according to thematic order, and structure according to methodological order. The most advanced and effective literature reviews are theme-based.

The outcomes of a literature review could be to write and submit a paper for an academic journal. Below are some recommendations for writing a literature review paper:

- Cover different years (historical development of the topic area).
- Cite many different authors.
- Establish links between the various research themes.
- Identify research gaps for future research.
- Explain how the outcomes contribute to the existing knowledge.

9.3.6   The Research Methods

The section on research methods is important because it contains not only the choice and actual methods used to solve research problem, but also sufficient detail for readers to replicate the work done and hopefully obtain similar results. Researchers should explain why the particular research method or methods is (are) used rather than other ones. Moreover, researchers need to clearly present the research process, such as data collection, data analysis, validation, and verification. We use two examples in this section to show how to write the research methods section. The two examples are from papers published in *Sustainable Cities and Society* (Alam et al. 2019). The paper titled 'Government championed strategies to overcome the barriers to public building energy efficiency retrofit projects' (https://doi.org/10.1016/j.scs.2018.09.022), aims to identify a comprehensive list of barriers to retrofitting public building stock for energy efficiency and associated strategies to address them (Example 9.10).

Example 9.10

The focused group method was used as the primary data collection method to solicit the perspectives of government officers on the barriers and coping strategies of implementing public building retrofitting projects (Bryman, 2016) [**Sentence 1**]. Focus groups were selected over other qualitative research methods because they can generate

---

(Continued)

(Continued)

information on the collective views of the selected participants regarding retrofitting barriers and the overcoming strategies [**Sentence 2**]. This method is useful in generating a rich understanding of participants' experiences and beliefs. Unlike interviews, where the researcher asks questions and controls the dynamics of the interview session, the dynamics of the focus group discussion is also influenced by the experience and interaction of the participants [**Sentence 3**]. This is particularly important in this research to understand the retrofitting barriers faced by the various participants that are positioned within certain divisions of the government organisation as well as barrier and strategy causality between divisions [**Sentence 4**].

In **Sentence 1**, the authors directly showed that focused group is the data collection method in this research. In **Sentences 2** and **3**, the authors described the strengths of focused group. Based on the characteristics of this study and the advantages of focused group, the authors elaborate on the reasons for using focused group in **Sentence 4**.

Example 9.11

Two focused groups were conducted in two states of Australia where public building retrofitting programs were not successful as stated in the previous section [**Sentence 1**]. The two state governments and participants involved in the focus groups are confidential [**Sentence 2**]. The criteria to select focus group participants were:

- Must be senior personnel (i.e. Director, Manager, Policy and Program Officer) in a government organisation [**Sentence 3**]; and
- Must have at least 10 years of experience in development, funding, implementation and management of building retrofitting projects [**Sentence 4**].

Following these criteria, two lists of participants were compiled with the help of government partners in those two states [**Sentence 5**]. As our focus was to understand the barriers and strategies from the perspective of the personnel employed within government departments or agencies, the focus groups included participants from the public sector only [**Sentence 6**]. The selected participants were invited through an email containing the scope and agenda of the focus group session [**Sentence 7**]. The number of participants was 10 and 28 for the first and second hosted focused groups, respectively [**Sentence 8**]. There are 90% of the participants in State Government A and 82% of the participants State Government B were Director, Manager and Policy and program officer [**Sentence 9**]. Hence, it can be considered that the personnel characteristics for each of the focus groups in these two states were similar [**Sentence 10**].

The focused group started with a half hour presentation from the research team to introduce the research topic, session objectives and the current best practices for retrofitting public buildings for energy efficiency [**Sentence 11**]. This was followed by a one and a half hour session including interactive thematic discussions [**Sentence 12**]. These discussions covered a wide range of retrofitting related topics including auditing, finance, procurement and mandates [**Sentence 13**]. During each thematic discussion session, each participant was requested to think about retrofit project barriers and coping strategies, from the perspective of their departments, as well as in the overall context of state government [**Sentence 14**]. The discussions were recorded and then transcribed for the analysis [**Sentence 15**].

The collected data were analysed using a thematic analysis approach [**Sentence 16**]. This is a qualitative data analysis method that seeks to identify patterned meaning across a dataset (Bryman, 2016) [**Sentence 17**]. The approach was particularly useful in this study since participants described similar barriers and coping strategies using different words [**Sentence 18**]. In this study, the themes of barriers and coping strategies were derived from a thorough reading of the transcripts of the discussions and the notes taken during the workshop [**Sentence 19**]. The topics that recurred more often were categorised as a theme using this analysis procedure [**Sentence 20**]. Repetition is one of the most common criteria for establishing the pattern; however, cautions were taken to make sure that identified themes are relevant to our research focus [**Sentence 21**]. The identified thematic barriers were then categorised depending on in which phase (e.g. building efficiency assessment, financing, procurement, etc.) of a retrofitting project they appear [**Sentence 22**]. NVivo software was used to map the interactions between the identified barrier themes and the phases [**Sentence 23**].

The data collected from the two focused groups were merged during the thematic analysis because both groups raised similar issues [**Sentence 24**]. The thematic analysis findings of the two focus groups were subsequently validated using an expert review panel consisting of two team leaders of public building retrofitting programs from two other state governments in Australia, as well as 'Head of Policy' from the Australian Energy Efficiency Council [**Sentence 25**]. The expert panel recommended the addition of only one additional barrier, which was the willingness of government to carry net-debt over the forward estimate period [**Sentence 26**]. This barrier is now included in the result section of this article [**Sentence 27**].

Example 9.11 illustrates the detailed steps of conducting focused group workshops. **Sentence 1** illustrates where focused group is implemented. In **Sentences 2–4**, the authors presented the criteria for selecting focus group participants. Following the proposed criteria, the authors showed the detailed process of participants' invitation in **Sentences 5–8**. In **Sentences 9** and **10**, the composition of the respondents was discussed.

The second paragraph details the process of implementing focused group where the length, topic, and way of discussion were presented. The third paragraph shows the process of data analysis. **Sentence 16** presents the method for data analysis (thematic analysis approach). In **Sentences 17** and **18**, the authors provided more detail about the method and showed it was the best choice. **Sentences 19–22** show the criteria of data analysis, and **Sentence 23** introduces the software used in data analysis. The third paragraph shows the validation of research findings.

9.3.7 The Results and Discussion

This section is about what the research found out, including direct results and discussion of the results. The results part aims to report the findings and answer the research questions. Researchers should present results concisely and logically. The results part can be organized around research objectives or hypotheses. It is useful to include visual elements, such as tables and diagrams, to present the results. Discussion is to interpret results in light of what is already known, and any new insight about the phenomenon or event being researched. This section should compare the results with previous studies and discuss the similarity, difference, and contribution to knowledge and theory. If unexpected results occurred, it is necessary to discuss them and evaluate their significance. Practical implications of the research outcomes and limitations should also be discussed.

9.3.8   The Conclusion

As the last section of an article, the conclusion, should be reciprocal with the introduction. The conclusion has the following functions:

- To remind the readers of the research aim and objectives.
- To summarize the important findings of the research.
- To evaluate its contribution to the development of theory.
- To recommend areas for future research.

To achieve the above functions, the conclusion normally has the following components:

- Restatement of aims and methodological approach of the research.
- Summary of findings.
- Evaluation of theoretical and practical contribution.
- Recommendations for future research.

Example 9.12 is from the paper 'Cloud-based safety information and communication system in infrastructure construction' published in *Safety Science* (Zou et al. 2017). For more details about the research and the publication, readers are directed to https://doi.org/10.1016/j.ssci.2017.05.006.

Example 9.12

In this research a cloud-based Infrastructure Safety Information Management System (MapSafe) has been designed and developed as a promising platform to improve the practice of safety management in infrastructure (e.g., road) construction projects [**Sentence 1**]. The Map-Safe has successfully integrated several currently available IT including cloud computing, GIS and mobile technology [**Sentence 2**]. The Map-Safe offers an electronic method for safety data/information collection and communication in construction sites [**Sentence 3**]. The captured safety data using MapSafe is stored and processed automatically in the cloud [**Sentence 4**]. The processed safety data/information can then be visualized on a map instantly to allow real-time safety decision making [**Sentence 5**]. The project safety supporting information such as design drawings in the pdf format and building information models can also be easily accessed through the MapSafe interface which is an advantage in the data management [**Sentence 6**]. Hence, MapSafe is effective as a safety information and communication management system for the construction industry [**Sentence 7**].

No doubt the construction industry will move into cloud computing based communication and management, sooner or later, and some first movers have already started this journey [**Sentence 8**]. The intellectual contribution and novelty of this research is that it points out to the way and maps out the processes for using cloud computing and free-to-use online web server for improving construction safety performance in real time communication, decision making and action taking [**Sentence 9**].

The current state of MapSafe has presented significant opportunities for future research and development in four areas: workflow automation of complete lifecycle management of various safety functions, real-time statistical analysis of individual safety activities, system security and data privacy, and web portal layout and functionality [**Sentence 10**]. In addition, in future follow-on research, data captured by this MapSafe system will be analysed to identify trends and patterns, which will help further improve construction safety [**Sentence 11**].

In the first paragraph, the authors revisited the cloud-based Infrastructure Safety Information Management System proposed in the research. **Sentences 1** and **2** indicate that the aim of this research has been fulfilled. In **Sentences 3–7** the functions and advantages of the system are described. They reflected that the proposed system can fill in the existing research gap. **Sentence 8** described the future trend of construction management research. In **Sentence 9**, the authors presented the contribution and novelty of the research. **Sentences 10** and **11** discussed future research based on the proposed system.

9.3.9 The Acknowledgement

If research is supported by some research funding, list the funding sources in the acknowledgment, e.g. 'the author wishes to express their sincere gratitude to [the funding name] (grant number: XXXX)'. In addition, researchers might express their gratitude to someone who helped them in the research, e.g. 'the author wishes to express their sincere gratitude to Dr XXX for his kind help in proofreading of an early version of the manuscript'.

9.3.10 The References

The importance of this section should not be underestimated, as it is a simple but straightforward way of checking if the research has included recent important literature and has covered a reasonable period of time as well as the key authors in the field. Different journals have different referencing styles. Endnote software can help researchers to automatically insert corresponding references into their manuscript.

9.3.11 Manuscript Component Checklist

Table 9.3 shows a checklist against the above-mentioned sections.

Table 9.3 Journal manuscript components.

Component name	Requirements
Title	Concise and attractive
Authors	List all authors who have contributed to this article
Abstract	Five sentences:
	The background of the research
	The importance of the research (i.e. why this research and why now)
	The research method used
	The main research findings (may be two sentences)
	The conclusion
Keywords	Up to five
Introduction	Provide a brief background of the research to show its importance and bring the readers' attention into the research and agree with the research aim
	Briefly describe the relevant critical theories
	Research aim and objectives
Literature Review	Comprehensive and up-to-date, use subheadings
	The Who – who has done similar research
	The Why – describe the nature and importance of the research,
	The What – What it is about
	The How: the methods used
	The Results: what have been found?
	The Summary: summarize the results
	Research gap and need: What needs to be researched
Research methods	Describe the methods used to obtain research data – survey? Interview? Case studies? etc. Why? Is it project based or organization/industry-wide?
	List the research parameters
	Data collection methods
	Data collection instruments
	Methods for analysing data
Results and discussion	What are the main findings and results? Are they similar to or different from existing studies and why?
	Verify or propose theoretical framework
Conclusion	What can be concluded, what are the limitations?
	Where to from here – the future research directions
	Originality/value (theoretical contribution /implication)
	Practical implications: what practice could do based on the research findings
Acknowledgement	Research funding
Key references	Covers main authors and historical development trends

9.4   Submission of Manuscript

Before submitting the manuscript to the target journal, researchers should carefully read and follow the 'Guides for Authors' provided by the journal. It is important to choose the right journal for submitting and publishing your research outcomes. Generally speaking, it is important to check the aims and scopes of the journal and analyse the recent volumes to make sure your research fits well with the journal scope and aims and aligns with its emerging trends of research directions.

9.4.1   Covering Letter

Editors are busy and they do not have time to read manuscripts in detail. The only way to understand manuscripts is the covering letter and maybe also the abstract and the manuscript headings. To a certain extent, a covering letter decides whether a manuscript is sent to reviewers or rejected at the editor's desktop. A complete covering letter should contain the following contents: (i) the novelty of this research; (ii) the contribution to the knowledge; (iii) the implementation of this research; (iv) the reason why this journal should publish this paper; (v) research background; (vi) research gap; (vii) research methods and processes; (viii) research findings; (ix) the statement about no conflicts of interest would affect the decision to publish the manuscript. An example of a cover letter is presented in Example 9.13.

Example 9.13

Date Here
Journal Name Here

Dear Editor in Chief,

On behalf of all co-authors, I would like to submit the enclosed manuscript entitled "Discovery of new safety knowledge from mining injury data in construction projects", which we wish to be considered for publication in your journal *Safety Science*. This research used several methods including statistical analysis, visualization analysis, and association rule mining (ARM) methods to mine a large injury dataset to discover the characteristics, patterns, causes, mechanisms of construction injury. Several important facts and new safety knowledge were obtained, and based on these findings, five strategies were developed to reduce injuries and improve construction safety performance.

   Data: Previous studies mainly used a small number of questionnaire surveys or interviews to obtain "cross-sectional" data at a particular point in time, and thus it is difficult to extract adequate knowledge and patterns for improving safety management, from a longitudinal perspective. This research used a large set of real injury and incident data provided by construction companies. The data were recorded at the time when injuries or incidents happened and covers a period of 10 years.

   Methods and Results: Several methods, including statistical analysis, visualization analysis, and associated rule mining methods were applied to analyse this large dataset. The characteristics of injury (e.g., time of accidents, bodily location of injury, nature of injury, and age of the injured), the causes and formation mechanism of injury were identified. Based on

---

(Continued)

(Continued)

the discovered mechanism, this study proposed five strategies for improving safety management performance, including pretask talk, personal protective equipment, on-site real-time monitoring, personalized safety training, and tailored first aid and medical preparedness.

Novelty: The novelty of this study includes providing a "panorama" of construction injury based on mining and analysing a large dataset with focus on the time of accidents, bodily location of injury, age of the injured, and characteristics, nature, and mechanism of the injury. Major causes of incidents and injuries are unfolded from this research. The knowledge discovered in this study could help practitioners better target their efforts to mitigate workplace injuries and prevent safety incidents. Meanwhile, researchers could benefit from gaining a better understanding of the characteristics and formation mechanism of injury, as well as how new research methods and data analysis methods may be applied to the field of safety science and management.

The **highlights** of this paper include:

- Identified overall pattern and trend of injury occurrence by months and time of the day.
- Identified and confirmed that the most injury-prone body parts, which are hand, foot, eye, and face.
- Derived 128 injury-causes rules.
- Analysed injury nature and causes with the corresponding body parts.
- Proposed five safety improvement strategies.

This research is innovative and informative and is within the scope of the journal and of interest to its readers.

We certify that we have participated sufficiently in the work to take public responsibility for the appropriateness of the collection, analysis, and interpretation of the data. This manuscript has not been submitted for publication elsewhere, and all the authors listed have approved the manuscript that is enclosed.

Thank you for your time and consideration. I look forward to hearing from you. If you have any queries, please do not hesitate to contact me.

Best regards.
Yours Sincerely,
Corresponding Author's Name Here

9.4.2 Highlights

Highlights is a short collection of bullet points that convey the core findings and provide readers with a quick textual overview of the article (ELSEVIER 2018). Usually, highlights contain 3–5 bullet points describing the essence of the research (e.g., research findings and conclusions). The wording and meaning of the highlight should be very condensed.

9.4.3 Title Page

The title page should contain title, author names and affiliations, corresponding author, address. An example of a title page is shown here (Example 9.14).

Example 9.14   Article title here (e.g., A mixed methods design for building occupants' energy behaviour research)

Author 1 name[*a], author 2 name [b], author 3 name [c], author 4 name [d]

   a * Corresponding author, Position title, organisation name. Email: xxx
   b Position title, organisation name. Email: xxx
   c Position title, organisation name. Email: xxx
   d Position title, organisation name. Email: xxx

9.4.4   Suggesting Referees

Some journals request authors to suggest several referees in the article submission system. Not only should the basic information about the referees be provided (e.g., name, email and institute) but also the reason for choosing them (Example 9.15).

Example 9.15

Suggested Reviewers: [Given name] [Surname]
   [Academic title], [Institute]
   [Email address]

   Reason: Prof XXX has expertise on the research of [the research field] and published many papers on the related topics regarding [the research field] in leading journals. His/Her suggestions to this journal would be valuable for enhancing the quality of this research paper.

9.4.5   Status of Submission

When researchers successfully submit a manuscript to a journal, they will enter into a waiting period for the manuscript to be reviewed. Different journals have different review periods. There are several statuses of the manuscript submission and the author services after manuscript acceptance, as shown in Table 9.4, which is cited from the Wiley-Blackwell website (Wiley, n.d.).

9.5   Responses to Reviewers' Comments

If the editor gives researchers a chance to revise the manuscript, whether it is major or minor revision, it means that the manuscript has a chance to be accepted. Researchers need to consider all comments, suggestions, and questions raised by the editor and reviewers, and revise the manuscript accordingly. Researchers need to submit detailed responses to the editor and reviewers' comments, and two copies of the revised manuscript – one with changes tracked, and the other a clean version. Line number is a method used to specify a

Table 9.4 Status of manuscript submission and author services.

Stage	Status	Description
Presubmission	New submission	The paper has been submitted successfully by the author and is waiting to be checked by the Managing Editor before being forwarded to the Editor.
	Awaiting allocation	The paper has been assigned to the Editor but has not yet been sent to reviewers (however, the Editor may have assigned the manuscript to an Associate Editor – there is no separate system status to indicate this). The Action Editor may decide to make a decision without allocating reviewers.
Review	Under review	The paper is with reviewers for comment or waiting for the Action Editor's decision. If the initial reviews are conflicting, the Action Editor may occasionally decide to approach an additional reviewer.
	Decision made – notification imminent	The Action Editor has written an email addressed to the corresponding author and the author will be notified of the decision as soon as the Managing Editor has proofread that email or the Editor has checked it.
Decision	Reject without review	The Action Editor has rejected the paper without sending it for peer review.
	Reject after review	The paper has been through the peer review process and the Action Editor has decided that it is not suitable for publication.
	Revise and resubmit	The paper has been through the peer review process and the Action Editor has decided that it may be suitable for publication after substantial changes are made. The Action Editor and reviewers will usually suggest improvements that will make the paper suitable for publication.
	Accept with minor amendments	The paper has been through the peer review process and the Action Editor has decided that it would be suitable for publication after some relatively minor changes.
	Accept	The Action Editor has decided that the paper is suitable for publication in its current form.
Author services	Accepted article received in production	The manuscript has been received by the typesetter for production to begin.
	Proofs sent	Typesetting of the proof has been completed, and an email alert with a link to the online proof has been sent to the corresponding author. Corrections should be returned as soon as possible.
	Corrections received	Author corrections have now been received. There will be a delay before your article appears online while the typesetter makes the corrections.
	EarlyView	The corrected article is now published online, ahead of inclusion in a print issue. You may view your article online at this stage. Please note that this is the final, published version of your article; no further changes can be made to it. You can download a PDF Offprint from Author Services.
	Issue published online	The issue containing the article is now published online. The print publication of the article in an issue may precede or follow this stage.

Source: Adapted from What Happens to My Paper? – The British Psychological Society.

particular sequence of characters or sentences in a manuscript. If the submission system cannot add the line numbers for the manuscript automatically, researchers need to add line number by themselves.

Example 9.16 presents responses to reviewers comments.

Example 9.16   Responses to the reviewers' comments

Paper title:
Paper ID:
Date of responses:

We appreciate the constructive comments and suggestions from editor and reviewers. We believe that there is a significant improvement in the quality of this manuscript through our efforts to revision. Point-by-point responses to the reviewer comments are provided as below:

Responses to the comments made by Reviewer #1:

Comments	Responses
Lines 72–74: Please discuss the method used in Lander's analysis.	Thank you for your suggestion. We have discussed the methods used in Lander's analysis as below: "Lander et al. (2016) analysed 23,464 work-related injuries from 1980 to 2010 to investigate injury trends according to age, severity, work activity and business cycle by using linear regression analysis. The annual incidences were calculated and employment levels were used as an indicator of fluctuations in the business cycle." Please see *Lines 86–89* in the revised manuscript.
Lines 52–54: Although I agree with this statement, it is best if you can provide a reference for it. Please see some relevant references including: https://doi.org/10.1080/01446193.2011.5525 12, https://doi.org/10.1016/j.jobe.2021.102398, https://doi.org/10.5130/ajceb.v13i2.3120	We appreciate your comments. We have provided references for the statement as follows (Please see *Line 55* in the revised manuscript): • Ajayi, S., Adegbenro, O., Alaka, H., Oyegoke, A., Manu, P., 2021. Addressing behavioural safety concerns on Qatari Mega projects. Journal of Building Engineering 41, 102398. • Hallowell, M., Esmaeili, B., Chinowsky, P., 2011. Safety risk interactions among highway construction work tasks. Construction Management and Economics 29, 417–429 • Panwar, A., Jha, K.N., 2021. Integrating Quality and Safety in Construction Scheduling Time-Cost Trade-Off Model. Journal of construction engineering and management 147, 04020160.
Line 353: please clarify the unit of measurement for the y-axis values in the different figures of Fig 8.	The Y-axis of Fig. 8 indicates the number of a particular type of injury as a proportion of the total number of injuries. Please see *Lines 351–352* in the revised manuscript.

(Continued)

(Continued)	
Comments	**Responses**
Table 1: Although it can be understood that the Asterix in the tables indicate statistically significant values, however, it will be best if you include it as a note below the table.	We have included the Asterix as a note below Table 1. Please see *Lines 376* in the revised manuscript.

9.6 Summary

Writing for publishing in an international peer-reviewed journal has become an aspirational goal and requirement for all researchers, particularly young and early career researchers and PhD students. While this chapter has provided the basic structure and contents of such technical research papers, and many templates and examples for different sections and contents are provided in this chapter, it is necessary for researchers to constantly practice writing. Writing is partly science and partly art. There are no fixed or standard templates, and authors may choose the best way of structuring and presenting their work. Do find out which journals best suit the manuscript, and follow the journal guidelines and requirements.

Responding to the comments made by the editor and reviewers of the manuscript is also partly science and partly art. The responses must be objective, polite, thorough, and to the point. Sufficient good reasons must be provided with carefully worded contents supported by evidence. Constant practising writing and improvement reflection is the golden rule of thumbs in achieving high quality publication.

Review Questions and Exercises

1 Which component is the most important in a manuscript for publishing in research journals?

2 How do you write an abstract?

3 How do you write an introduction?

4 How do you undertake a literature review?

5 How do you write a literature review-based paper?

6 What is the best way to present the research results and discussion?

7 What contents and components should be included in the conclusion section?

8 How many references should be included in a manuscript?

9 Why is it important to follow the Guides for Authors set by the target journal?

References

Alam, M., Zou, P.X.W., Stewart, R. et al. (2019). Government championed strategies to overcome the barriers to public building energy efficiency retrofit projects. *Sustainable Cities and Society* 44: 56–69.

Booth, A., Papaioannou, D., and Sutton, A. (2012). *Systematic Approaches to the Literature.* London: Sage.

ELSEVIER (2018) Highlights. https://www.elsevier.com/authors/tools-and-resources/highlights.

Hilary, G. (2010). *Science Research Writing: A Guide for Non-Native Speakers of English.* London: Imperial College Press.

Li, Z., Shen, G.Q., and Xue, X. (2014). Critical review of the research on the management of prefabricated construction. *Habitat International* 43: 240–249.

Lu, W. and Yuan, H. (2011). A framework for understanding waste management studies in construction. *Waste Management* 31 (6): 1252–1260.

Niu, S.Y., Pan, W., and Zhao, Y.S. (2016). A virtual reality integrated design approach to improving occupancy information integrity for closing the building energy performance gap. *Sustainable Cities and Society* 27: 275–286.

Siegel, R.D. (2020) Reading scientific papers. https://web.stanford.edu/~siegelr/readingsci.htm; accessed 9 April 2023.

Wiley (n.d.) What happens to my paper? https://onlinelibrary.wiley.com/pb-assets/assets/20448295/What_Happens_to_My_Paper.pdf; accessed 9 April 2023.

Xu, X.X. and Zou, P.X.W. (2020). Analysis of factors and their hierarchical relationships influencing building energy performance using interpretive structural modelling (ISM) approach. *Journal of Cleaner Production* 272: 122650.

Xu, X.X. and Zou, P.X.W. (2021). Discovery of new safety knowledge from mining large injury dataset in construction. *Safety Science* 144: 105481.

Zou, P.X.W., Lun, P., Cipolla, D., and Mohamed, S. (2017). Cloud-based safety information and communication system in infrastructure construction. *Safety Science* 98: 50–69.

Zou, P.X.W., Xu, X.X., Sanjayan, J., and Wang, J.Y. (2018a). A mixed methods design for building occupants' energy behavior research. *Energy and Buildings* 166: 239–249.

Zou, P.X.W., Xu, X.X., Sanjayan, J., and Wang, J.Y. (2018b). Review of 10 years research on building energy performance gap: life-cycle and stakeholder perspectives. *Energy and Buildings* 178: 165–181.

10

Thesis Writing

10.1 Introduction

In the previous chapters, the fundamentals, structure, and components of research and various research methods and journal manuscript writing have been discussed. This chapter focuses on thesis writing, including the overall design, components, structures, innovation, theory development, and contribution to knowledge. It also provides insight into the criteria and process of thesis examination. This chapter is specifically written for PhD and Masters research students, while it is also a good reference for final-year undergraduate honours degree students' thesis writing. The PhD thesis, in particular, is research work that requires a certain degree of originality and contribution to theory and discovery of new knowledge.

10.2 Overall Design of the Thesis

Research design plays an important role in managing the research process. Sharpening your axe will not delay your job of chopping wood. A good research design can help researchers yield twice the results with half the effort. In general, research design involves defining the aims and contents of the research and determining the best way and process to conduct the research. Figure 10.1 shows the life cycle of a thesis project for obtaining a research degree, starting from research problem definition and research aim and objectives specifications to thesis submission and examination.

The first and most important step in research design is to find the research problem. Research problems can be derived from observation of a social or natural phenomenon combined with relevant literature and theory, from personal or practical experience, or from extensive reading of the literature. Generally speaking, junior researchers obtain their research problems mostly through reading the literature and identifying gaps in existing knowledge. Once the research problems have been identified, they need to be further specified and transformed into research aims and questions. The number of specific research questions should be limited to three to five. In the process of formulating the research questions the following points should be considered: (i) Are the research questions of value in solving in practice? (ii) Do the research questions have theoretical value? (iii) Are the scopes of the research questions clearly defined? (iv) Are the research questions feasible to solve?

Research Methodology and Strategy: Theory and Practice, First Edition. Patrick X.W. Zou and Xiaoxiao Xu.
© 2023 John Wiley & Sons Ltd. Published 2023 by John Wiley & Sons Ltd.

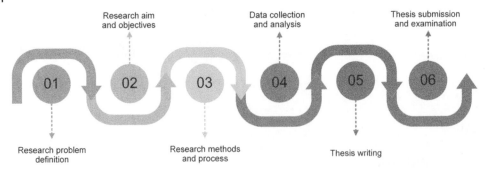

Figure 10.1 Overall design and process of a thesis research project.

When designing, researchers should be clear about the purpose of their research. Researchers need to use the aims of the research to explain why a particular study is being done and what the research is trying to achieve. These research aims and objectives should be closely linked to the research problems. Once the research problems and objectives are clear, the researcher needs to provide a theoretical basis for the research. The presentation and discussion of a theoretical basis not only argue for the validity of the research but also demonstrate that the research is worthwhile.

All research requires a rigorous and detailed research method. Research methods should be consistent with the research problem and research objectives. The selection of research methods should follow the principle of appropriateness: only methods that are applicable to solving the proposed research problem are good methods.

Once the research methods and processes are clear, data need to be collected. Researchers should ensure that the data collection process is harmless to the research subjects and should follow research ethics. Data collection methods include but are not limited to experiments, questionnaire surveys, and interviews. Data may also be obtained from government statistical bureaus or data management agencies and companies or other reliable resources. Once the data have been collected, they are processed and analysed to draw conclusions to achieve research aims and objectives. The results of the data analysis should also link with relevant theories. Figure 10.2 shows the typical research process of PhD research, including research activities.

Figure 10.2 Research process of a PhD thesis.

10.3   Innovation of the Research

Innovation is the soul of an academic research thesis. Research students should focus their thinking on three questions around innovation, as discussed in the following sections. Only when there is innovation will the value be reflected and results recognized by others.

1) **What is innovation?** There are three ways in which students can present and discuss the innovation of thesis: (i) innovation of research problem; (ii) innovation of research method; and (iii) innovation of theory. A research problem is innovative if it has not yet been proposed but has significant research implications. Innovation of research method can involve proposing a new method or introducing a research method from other fields to solve a specific research problem. When a researcher proposes a new theory or extends an existing theory, this is considered theoretical innovation.

2) **Does the innovation proposed make sense?** Research students are expected to present the theoretical and practical context in which the innovation is proposed. It is important to convince not only themselves but also the readers and the examiners that the proposed innovation makes sense to proceed with, that is it is indeed of theoretical and practical importance and worthy of time and effort to read.

3) **How to prove the innovation proposed is valid?** Research students need to provide detailed evidence that the proposed innovation is credible. This would mean: detailing that the problem proposed is one that has not been considered before and does have research implications; arguing that innovative research methods are appropriate and have not been used before; and/or demonstrating new theoretical insights and foresight into the nature, laws, and trends of development in the chosen research field.

10.4   Structure of a Thesis

10.4.1   The Title, Abstract, and Keywords

The title of a thesis is crucial, as it tells the readers what the thesis is about. It could be a new innovative perspective, it could be introducing a new method, it could be defining and solving a new research problem, or it could be a combination of the above. The thesis title should be catchy, concise, and reflective of the thesis, usually containing several keywords of the thesis. In a way it is similar to the title for a journal manuscript, but slightly longer and broader.

An abstract for a thesis is often contained within one A4 page. It is not easy to be concise in this one-page limit. It is equal to a mini version of the whole thesis. Therefore, it is important to put effort into it. Most examiners read the abstract first before other parts of the thesis, which means it is the first impression of the thesis and it counts.

Generally, the first paragraph introduces the research background, including the importance and significance of the research and the state-of-art of the problems being investigated, from both practical and theoretical perspectives. There are normally three to four sentences in this paragraph. Following the opening paragraph comes the research aim and objectives. This is the most important section and defines what the research is about. One sentence describes the overall aim, followed by four to five sentences describing the research objectives in a manner that is interrelated and logically linked from one objective to the second, the third, and so on.

The third section should be the research methods used, which could be a qualitative, quantitative, or mixed method, or a data-driven method, or most likely a combination of these methods. The methods for collecting and analysing any data also need to be described. The process of data collection is key to achieving meaningful results. If the methods and processes of data collection are incorrect, the data collected may not be representative or may be biased or subjective. The research may become 'rubbish in and rubbish out'. Even with good quality data collected, if the analytical methods are not suitable or inappropriate, no meaningful results could be obtained. It is for this reason that this book has provided detailed discussions on the data collection and analysis methods. If data are not representative, not objective, the analysis would be meaningless, with a knock-on effect for the results and discussions and so on. It is so important to think about and provide details how data are collected and analysed whilst detailing the research methods.

The fourth section is an overview of results, which naturally correspond to the research objectives, and brief discussion about the results.

The final section is a conclusion alongside implications of the research, from both the theoretical perspective and practical perspectives.

10.4.2 The Introduction

This section addresses the importance and significance as well as the urgency of the research. It opens up a story and tells the readers, *this is an important story*. An introduction should include the background, the current state of the relevant literature, the significance of the research, the research problems, the research aim and objectives, as well as thesis structure and layout. It is necessary to ground the research in context, which may include introducing theory. Although the introductory chapter may not be long, its importance should not be underestimated.

The introduction should make readers believe and agree with what the research aims to do, and its relevance to the society and/or daily life. A major factor in the introductory chapter is the use of a logical structure and convincing statements. If the readers are not convinced after reading the introduction, the rest becomes less credible.

Research students may use a big picture as the opening statement at the starting point, then zoom into specific practical problems, which can then be turned into research aim and objectives. For example, in the current political, socio-economical, and environmental context, carbon dioxide emission reduction and energy saving are big issues worldwide. The research could include such points in the introductory section, backed by statements from the United Nations (UN) or other well recognized international organizations. Likewise, if a thesis research is about safety, the introduction might start with a statement that says human life is a globally important issue and safety is about life or death, therefore it is important to continue to undertake research into improving safety conditions, safety behaviour, and safety performances in professional jobs and daily life.

The general structure of an introduction is:

- Significance and importance
- Research problem statement
- Research aims and objectives

- Research questions
- Scope and context
- Research methods
- Thesis structure

10.4.3 The Literature Review, Theory, and Hypothesis

Every research should include a review of the literature to understand the state-of-the-art of the research in the field. Only by reviewing the literature can the research gaps and needs be identified, the research aim defined, and the innovation and contribution to the theory and knowledge feasible. There are different methods of undertaking literature review. The simple method is to list in chronical order the authors who have undertaken research in the topic area. A better method would be to follow the research objectives and review and group and discuss the relevant literature according to these research objectives. This requires logical evaluation, assessment and analysis of the literature. Section 9.3.5 of Chapter 9 provides details on how to undertake a literature review.

A thesis should be built on the understanding of the relevant theories related to the research topic. In PhD research, it would be appropriate to draw on several relevant theories from different domains and disciplines, but this is dependent on the nature of the research problems, scope, and complications of these problems. If interdisciplinary perspectives are used then it would be useful to look at the relevant theories from different disciplines. It is also necessary to review different research methods that have been used in the literature.

After a comprehensive and critical review of the literature and relevant theories, it comes to the development of research questions to be answered or research hypothesis to be tested. The research hypothesis needs to have independent variables and dependent variables. In between these two variables are the control variables and mediator variables.

10.4.4 The Research Methods, Design, and Process

This section requires a comprehensive research plan that includes step-by-step processes and activities to achieve the research aim and objectives. It needs to discuss the method or a combination of different methods to be used for collecting data and assessing data saturation and data collection completion. It also needs to provide details on the data analysis methods and how these methods are selected. There has been discussion about research methodologies in this book. Researchers could refer to the relevant sections to understand different methods and their strengths and weaknesses and advantages and disadvantages in order to select the most suitable research methods.

10.4.5 The Results

This section is crucial and should be corresponding to the research aim and objectives. This means that if there are five research objectives, then there should be five sections of the results to demonstrate the achievement of the objectives as the outcomes of the research. The results should be based on the thorough analysis of the research data that was properly collected. The description of the results should be as detailed and thorough as possible. It is effective to use subheadings to present the results in a structured format. Tables and figures should be used to presentthe results.

10.4.6 The Discussion

This section is perhaps the most difficult section to write, as it requires research students to discuss their research results in relation to the literature, current knowledge base, and relevant practice. It is important to discuss the results with as much detail as possible in relation to the current knowledge and theory. Surprising results could become the highlight of the research. If the research results are not in alignment with the current understanding of the phenomena, then the discussions should provide explanations and reasons. If something cannot be explained in detail, then it should be suggested for future research.

It will be helpful to develop a theoretical framework, to summarize the results and discussions, toward the end of the discussion section. This framework becomes part of the contribution of this research into the current theory or knowledge base. The discussion could also draw onto the innovation of the research and implication to practice.

10.4.7 The Conclusion

This section should be different from the introduction. It should really conclude something, a takeaway message. In many cases, the generalization of the research results is presented in the conclusion section. The conclusion also needs to present the innovations of the study. The contribution of the research should be highlighted in the conclusion. Contributions are generally divided into theoretical and practical aspects. Theoretical contribution is mainly reflected in the theory development and advancement in existing theory, in one or more of the following ways:

- Does the research add new elements to a theory?
- Does the research change the internal logic of a theory?
- Does the research broaden the application of a theory?

The practical contribution focuses on whether the findings and outcomes can be applied in practice. Research students may think about who their research is for, how the outcomes of the research can help them, and what practical benefits can be achieved. For example, if the research makes a policy recommendation, research students could elaborate on who their policy recommendation could be used for, what effects and benefits it may generate, and what practical problems it could solve.

The conclusion should also discuss the limitations of the research and what the future research in the same topic area should focus on. Remember that no new information should be included in the conclusion section.

10.4.8 Other Components

The thesis may include other components, such as an acknowledgement, ethical statement, appendix, etc. The key is to keep it simple and maintain truth and completeness. In the acknowledgement section, some students may extend their thanks and gratitude to their friends and families, while these are not wrong, there are arguments for not including this types of content, as the thesis is a piece of academic research work. There are also suggestions that the acknowledgement could include sentences thanking previous researchers whose publications have been cited in the current research thesis.

It is suggested that research students read a few recent good PhD or Master theses to get an understanding of the general requirements, format, and the amount of work for a good PhD or Master thesis. These good theses could also serve as benchmarks for your own writing of the thesis. It is important to find the best theses in the same field of research around the world for learning and benchmarking.

10.5 Thesis Examination Criteria

Research students should be clear about the process and criteria for the thesis examination. PhD and Master theses should be specialized, theoretical, and innovative. Specialization means that the thesis is an in-depth study of a specialized subject area and meets the requirements of the specialized subject. Theoretical means that the thesis should be based on certain theories to analyse and discuss the research object, and to generalize new theories through an in-depth analysis of the research object. Innovation is something that has not been done before, or has been studied but not in sufficient depth and needs to be enriched and developed. Innovation is the soul of the thesis and an important criterion for judging its quality.

The thesis should also demonstrate a substantial and original contribution to knowledge, in the form of new knowledge or significant and original adaption, or application and interpretation of existing knowledge. Thesis examiners generally ask the following questions when examining a thesis:

- Does the thesis makes an original contribution to the knowledge of the subject?
- Does the thesis reflects a solid and broad theoretical foundation and systematic and in-depth expertise in the discipline and related fields?
- Does the thesis provides a comprehensive and systematic literature review?
- Does the thesis adopt appropriate methods?
- Are the research findings suitably set out, accompanied by adequate exposition, and discussed critically?
- Is the quality of language and general presentation satisfactory?

It is worthwhile keeping these questions in mind when writing and revising the thesis.

A PhD thesis is usually sent to three examiners who are experts in the same research field. A Master's thesis is normally sent to two examiners. The examination results could fall into one of the five categories: pass without correction needed, minor revision subject to the approval of the supervisor, major revision subject to the approval of the research committee, major revision, and re-examination and fail. In most cases, students are required to revise the thesis according to these examination outcomes.

Several universities' PhD thesis examination criteria are shown in Table 10.1, as examples. It can be seen from Table 10.1 that universities have strict thesis requirements on 'originality of research', 'scientific nature of the research questions', 'comprehensiveness of

Table 10.1 Examination criteria of PhD thesis.

University	Evaluation criteria
Cornell University (https://gradschool.cornell.edu/about/program-metrics-assessments-and-outcomes/learning-assessment)	A candidate for a doctoral degree is expected to demonstrate *mastery of knowledge* in the chosen discipline and to *synthesize and create new knowledge, making an original and substantial contribution* to the discipline in an appropriate timeframe. a) Make an *original and substantial contribution* to the discipline. Think originally and independently to develop concepts and methodologies; Identify new research opportunities within one's field. b) Demonstrate *advanced research skill*. Synthesize existing knowledge, identifying and accessing appropriate resources and other sources of relevant information and critically analysing and evaluating one's own findings and those of others; Master application of existing research methodologies, techniques, and technical skills; Communicate in a style appropriate to the discipline. c) Demonstrate commitment to advancing the *values of scholarship*. Keep abreast of current advances within one's field and related areas; Show commitment to personal professional development through engagement in professional societies, publication, and other knowledge transfer modes; Show a commitment to creating an environment that supports learning through teaching, collaborative inquiry, mentoring, or demonstration. d) Demonstrate *professional skills. Adhere to ethical standards in the discipline; Listen, give, and receive feedback effectively*
University of Melbourne (https://policy.unimelb.edu.au/MPF1321)	Examiners must consider the thesis or compilation solely on its merits and must consider whether it meets the following criteria: a) the candidate has demonstrated sufficient familiarity with, and understanding and *critical appraisal of*, the relevant literature; b) it is a sufficiently comprehensive investigation of the topic; c) methods and techniques adopted are *appropriate, properly justified and applied*; d) results are suitably set out and accompanied by *adequate exposition and interpretation*; e) conclusions and implications are appropriately developed and clearly *linked to the nature and content of the research framework and findings*; f) *research questions have been tested or explored* according to disciplinary norms; g) literary quality and general presentation of the thesis is of an appropriately high standard; and h) the thesis or compilation as a whole constitutes an *original contribution* to knowledge in its subject area.
University of New South Wales (https://www.unsw.edu.au/content/dam/pdfs/governance/policy/2022-01-policies/thesisexamproc.pdf)	Examiners are expected to submit to the University a recommendation (as detailed in the relevant Conditions of Award Policy) regarding the thesis and to provide a written report on the thesis that provides a strong justification for their recommendation. Where indicated, the examiner must provide guidance to the candidate regarding any changes required. The examiners are asked to examine the thesis against the following criteria: a) Does the candidate demonstrate a significant and *original contribution to knowledge* (relative to the level of the degree being sought)? b) Does the candidate engage with the literature and the work of others? c) Does the candidate show an *advanced knowledge of research principles and methods* related to the applicable discipline?

Table 10.1 (Continued)

University	Evaluation criteria
ETH Zurich (https://ethz.ch/content/dam/ethz/special-interest/phys/department/doctoral/Info%20Sheet%20Report%20DPHYS%20August16.pdf)	To judge the *quality and the innovative contribution* of a doctoral thesis, the following items should be considered: a) Form and structure • Is the work well structured? Is the structure logical? • How good is the level of language? • Is the format of the text, graphics, and tables adequate? b) Introduction and goals • Is the introduction adequate in length and detail? • Are the research questions and goals clearly stated? c) Methods • Are the methods clearly described and is their use justified? d) Results • Is the analysis of the data and results clearly described and is their interpretation conclusive and justified? e) Discussion and conclusions • Does the work contain a critical discussion of the methods and results? • Are the results and interpretation compared with the available literature? Is the own research achievement clearly distinguished from other scientists' work? • Is there a justification for the conclusions based on the results? • Does the doctoral thesis provide an outlook or a perspective of the field? f) Literature • Is the cited literature adequate/relevant and complete?
University of Toronto (https://www.sgs.utoronto.ca/current-students/program-completion/producing-your-thesis/doctoral-thesis-guidelines/student-guidelines-for-the-doctoral-thesis)	Regardless of the format of the doctoral thesis, certain criteria must be met as below: a) Demonstrate how your research makes an *original contribution* by *advancing knowledge* in your field. b) Show a thorough familiarity with the field and an ability to *critically analyse the relevant literature*. c) Display a mastery of research methods and their application. d) Offer a complete and systematic account of your scholarly work. e) Present the results and analysis of your original research. f) Document your sources and support your claims. g) Locate your work *within the broader field or discipline*. h) Write in a style that respects the *norms of academic* and *scholarly communication*.

literature review', 'rationale of research methods and design', 'significance of research contributions', and 'logic of the structure'. While this book provides higher degree research students with systematic guidance according to these requirements, different disciplines may have specific requirements, which students should also read and follow.

10.6 Pitfalls to Be Avoided in Thesis Researching and Writing

Thesis writing is a major project that requires a lot of effort from the students. Many students may encounter various pitfalls in the process of writing their thesis. Below is a list of possible pitfalls to be avoided to make the process less difficult.

Pitfall 1: Students might simply think that their thesis is innovative if it is done in a way that no one else has done. It is worth noting that what others have not studied does not necessarily mean innovative, instead it could be unworthy research. Students should also focus on the significance of the research when choosing a topic and defining the research problems.

Pitfall 2: Students might hold the view that only a 'big' topic can reflect the theoretical nature of a thesis and demonstrate the theoretical knowledge and sophisticated research skills. The research topic and problems for a thesis should be specific with sufficient depth.

Pitfall 3: Students might think that it is okay to 'borrow' sentences from other articles and papers or other types of references. This is plagiarism. Where the work of others is referenced, the original source should be cited.

Pitfall 4: Students might think that the thesis could be simply combining published journal articles into one piece. Students should pay attention to the logic of a thesis structure when combining their previously published journal papers into a thesis.

Pitfall 5: Students might think that a thesis must use 'high-end' research methods to highlight the innovation. It is not always better to use a method that looks more advanced. Only those methods that are suitable for solving the research problem proposed are good methods.

Pitfall 6: Students might write their literature review by simply listing the literature together. The worst literature review is all list without discussion. Students should have a logical framework to categorize and examine the literature.

Pitfall 7: Students might just re-describe the results in a different way instead of discussing them, in the discussion section. Discussion of the research results requires a high-level thinking and summarization. Discussion should include contents comparing the results with those in the current literature. It is useful to conceptualize the research findings into some sort of framework at the end of the discussion section.

10.7 Accessing and Reading past PhD Thesis

During the PhD research process, particularly at the beginning stage, it will be useful and helpful to read some good past PhD theses, to obtain an overall understanding of the structures and contents of a PhD thesis. Table 10.2 is a list of the channels and methods for accessing past PhD theses. In addition to this list, students should also search for past PhD theses in their university libraries. Note that PhD theses may not be released to the public until three years after the completion. Students must focus on the good PhD theses, as they will serve as benchmarks. When reading past PhD theses, students should learn about not only the literature review and research methods but also the structure, the logic, the methods for developing theory, and the style of writing.

Table 10.2 Channels and methods for obtaining past PhD theses.

Channels	Description	Website
ProQuest	ProQuest is the world's largest and most widely used database of theses. It covers the fields of science, engineering, economics and management sciences, health and medicine, history, humanities and social sciences. Through university library platform	https://www.proquest.com
OATD	OATD is a website focused on collecting theses and a repository for finding published open access theses from around the world. It contains over six million theses from over 1100 universities and research institutions worldwide. Open access	https://www.oatd.org
DART-Europe	DART-Europe is a collaboration of research libraries and library consortia dedicated to improving global access to European research papers, offering open access research papers from 572 universities in 29 European countries, currently aggregating 1159 476 resources, open access and free to download, read and use for all users Open access	https://www.dart-europe.org/basic-search.php
NDLTD	NDLTD is an international organization that through leadership and innovation, promotes the adoption, creation, use, dissemination, and preservation of electronic theses and dissertations. The NDLTD encourages and supports the efforts of institutes of higher education and their communities to develop electronic publishing and digital libraries (including repositories), thus enabling them to share knowledge more effectively in order to unlock the potential benefits worldwide. Open access	https://ndltd.org
University library	Many university libraries provide access to past PhD theses around the world.	

10.8 Summary

This chapter is written particularly for research students, including PhD, Master's and final year undergraduate honours degree students who are required to undertake research and write a thesis or dissertation as part of their academic study processes and requirements. It first discussed the overall design of a thesis, which includes research problem definition, research aim and objectives development, research methods selection and research process design, research data collection and analysis, thesis writing, submission and examination process, and criteria. The chapter further discussed research innovation, which is the soul of a thesis and plays a vitally important role and may include one or more aspects: innovation of research problem, innovation of research methods, and innovation of theory.

This chapter then provided detailed discussions on the structure of a thesis, which includes: title, abstract, keywords, introduction, literature review, theory, hypothesis, research methods, design and process, results and discussion, conclusion, implication, and other necessary components. The thesis examination criteria and process, and some pitfalls are also discussed in detail, which will be particularly useful to research students, as they might be lacking knowledge and experience in completing these tasks in their research journey. We suggest that research students who need to complete a thesis pay attention to the contents presented in this chapter and put effort to answering the review questions below.

Review Questions and Exercises

1 Why is innovation important in a higher degree research thesis?

2 How do you ensure that the thesis has an innovation and achieves the innovation requirement?

3 What is your understanding of theoretical foundation and theoretical contribution in higher degree research thesis?

4 How do you structure a thesis?

5 What are the thesis examination criteria?

6 What are the pitfalls in undertaking PhD thesis research? How do you avoid the pitfalls?

7 What are the differences between a higher degree research thesis and journal manuscript?

11

Research–Practice Nexus and Knowledge Coproduction

11.1 Introduction

The days of conducting research in an ivory tower then pushing the research results out for other academics and industry practical utilization are gone. Instead, new modes of collaboration are being developed and implemented, for example the industry–university partnership mode and the industry-lead mode (Berbegal-Mirabent et al. 2015; Moses et al. 2008). Figure 11.1 shows an example of universities and research institutes involved in the process of developing pile construction plans of a major bridge, through joint research and development. With such changes, it is important for researchers to shift their mindsets and change their operational processes to adapt to this new mode of operation. More importantly, researchers should recognize that it is through this mode of partnering collaborative operation that new knowledge is coproduced and applied to improve practice and contribute to new theory development in a seamless way (Pizmony-Levy et al. 2021). There have been some of such research modes at national and state governments' strategic levels for funding allocations, in many countries including the United States, United Kingdom, Australia, and China (Kerner 2008; Kwon 2022). However, academic researchers and industry practitioners are lacking the knowledge and experience to effectively implement such new modes of research–practice nexus. This chapter provides discussions and a framework to address this issue. The framework was developed based on literature review and case studies of collaborative research projects with government departments and agencies and leading industrial organizations.

11.2 Research–Practice Gap and Nexus

Research and practice have been separated under the influence of actors, practice, expertise, organizational contexts, and interests; this becomes a research–practice gap (Ye 2021). The research–practice gap is defined by the absence of reciprocal communication between the research and practice communities and limited adoption of evidence-based interventions in practice settings (Neal et al. 2015).

A strong research–practice nexus can serve to legitimize researchers' academic pursuits by addressing research problems that contribute to the resolution of practical challenges

Research Methodology and Strategy: Theory and Practice, First Edition. Patrick X.W. Zou and Xiaoxiao Xu.
© 2023 John Wiley & Sons Ltd. Published 2023 by John Wiley & Sons Ltd.

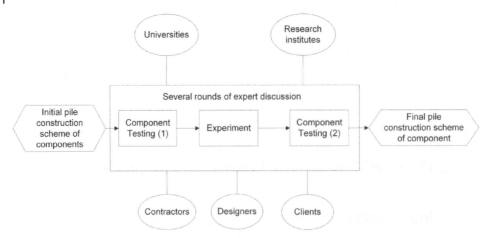

Figure 11.1 Example of university–industry collaborative research to develop bridge pile construction plan.

(Podgorodnichenko et al. 2022). Meanwhile, academic research may help practitioners understand the performance implications of decision making and management, by explaining, characterizing, developing, and enhancing practice, and therefore play a potentially essential role in guiding the creation of new practices that fit evolving practical demands (Unerman and O'Dwyer 2010).

To bridge the research–practice gap, there had been some research conducted on the research–practice nexus. For instance, Zou and Sunindijo (2015) proposed a seven-step cyclic iteration and nexus between research, theory, and practice, which is a mixed-methods research design. The seven steps are: (i) conceptualization, (ii) confirmation, (iii) theory development, (iv) action research, (v) modification, (vi) training and integration, and (vii) generalization of fresh theoretical positions. Neal et al. (2015) pointed out that the research–practice gap can be narrowed by enhancing researchers' dissemination efforts and practitioners' information searches. Makin (2021) found that there are signs that the research–practice gap can be bridged when academics and practitioners collaborate to foster deeper involvement and when effective boundary-spanning items are deployed. These are general steps, general views, and guiding points. More detail and specific information are needed to operationalize the concept.

11.3 Assessing the Research–Practice Nexus

Previous studies provided a checklist for assessing the relevance of research outcomes in practical application, which included five aspects: (i) contribution of the research to practice, (ii) theoretical approach, (iii) methodology, (iv) data analysis and synthesis, and (v) self-conscious integration of theory and practice (Evans 2009; Zou and Sunindijo 2015). Referring to each of these points, some details are provided here.

1) **Contribution of research to practice**: this is about identifying the measures that show the potential of the research for improving existing practice, and verifying the sources of data and tools used to measure practical performance as well as reflecting on the relevance of the research outcomes for improving practice (Böckel et al. 2021).

2) **Theoretical approach**: this is about making sure that the theory or approach can be verified, and that the theoretical approach and propositions can be justified; as well as reflecting on potential bias in the theory and amendments to the theory to make sound contribution to knowledge claims (Bolander et al. 2021).

3) **Methodology**: this is about ascertaining that the methodology allows for the verification of the theory and the methods are tried and trusted, and using appropriate methodology and reliability and validity measurement tools, as well as reflecting on the inherent bias in the method and the data collection process.

4) **Data analysis and synthesis**: this is about maintaining the credibility of results and generating sufficient evidence to support the results, and verifying the data using a combination of different methods (such as triangulation), as well as reflecting on the reliability and generalizability of the analysis and results that are applicable in practice.

5) **Self-conscious integration of theory and practice**: this is about identifying research elements that are relevant to practice, identifying missing research elements, and making sure that research results are communicated and accessible to industry practitioners (Joyce and Cartwright 2020).

11.4   Knowledge Coproduction

Researchers need to move one step back and do more fundamental work by exploring how knowledge is constructed in the first place (Tsoukas and Mylonopoulos 2004). Workplace traditions have a significant role in knowledge and skill development (Baarts 2009). A lot of learning and practice improvement takes place in the social world, among and through other people. Therefore, the integration of the realms of theory and practice is needed to ensure that research findings are relevant to the promotion of practice improvement (Zou and Sunindijo 2015). Knowledge coproduction is suitable for addressing this problem.

11.4.1   Knowledge Coproduction Definitions and Scopes

Knowledge coproduction is based on the interactions and dynamics between academic researchers and nonacademic practitioners. It promises to address the complex nature of contemporary challenges better than traditional scientific approaches and it has attracted more and more attention as an innovative solution to bridging the research-to-practice gap (Boaz et al. 2018). Conceptually, knowledge coproduction is part of a loosely linked and evolving cluster of participatory and transdisciplinary research approaches that have emerged in recent decades. These approaches reject the notion that scientists alone identify the issue, research the problem, and then deliver knowledge to society, in favour of more interactive arrangements between academic and nonacademic actors (Norström et al. 2020). Knowledge coproduction has moved from a niche area of scientific practice to the mainstream (Editorial 2018). Knowledge coproduction approaches include participatory research, mode-2 science, interactive research, civic science, postnormal science, transdisciplinary and joint knowledge production, action research, translational ecology, and engaged scholarship; these have become increasingly prominent during the past four decades (Norström et al. 2020).

11.4.2 Knowledge Coproduction Principles and Processes

Knowledge coproduction is context-driven, problem-focused and requires the engagement of multiple disciplines, and there are different ways of conceptualizing and implementing knowledge coproduction. Norström et al. (2020) provided a brief history of the development of the concept of knowledge coproduction from a sustainability research perspective. Norström et al. (2020) identified four principles of knowledge coproduction and presented a set of considerations for implementing and monitoring how the four principles are put into practice (Table 11.1).

Table 11.1 Knowledge coproduction principles and processes.

Principle name	Principle meaning	Processes
Context-based	• Situate the process in a particular context, place or issue. • Describe the coproduction process in contextually relevant language and based on a shared understanding of key concepts and terminology.	• Asking questions about how a particular challenge has emerged and how changing circumstances are likely to influence the work. • Who will be impacted or affected by the process and its outcomes. • Who has the power to enable or constrain action; how will policy, regulatory, institutional and cultural factors shape the process and the realization of desired outcomes.
Pluralistic	• Explicitly recognize the multiple ways of knowing and doing. All knowledge is inevitably situated and partial. Ensure a range of perspectives are included.	• Bring together academics and people from other sectors – government, business, civil society, local and indigenous communities, to generate knowledge and catalyse change; include different genders, ethnicity, age and nationality.
Goal-oriented	• Articulate clearly defined, shared and meaningful goals that are related to the challenge at hand; • Be problem focused and outcome focused. • Focus on needs and expectations.	• Identify goals of individual participants. • Establish overarching goals, including measurable goals and unmeasurable goals. • Recognize goals may evolve with time and the progress of the research. • Set meaningful and achievable milestones. • Undertake risk assessment. • Optimize pathways to reach the agreed goals.
Interactive	• Allow for ongoing learning among actors, active engagement, and frequent interactions. • Share knowledge, experience, ideas, and values. • Incorporate the knowledge coproduced in decision making; • Focuses on credible, salient, and legitimate.	• Knowledge coproduction requires frequent interactions among participants to occur throughout the process. • Interactions happen at all stages of the research project – framing and designing the research agenda, conducting the research, and using and disseminating the knowledge generated. • Develop trust among participants through dialogues. • Create active engagement and conversations. • Emphases on ongoing learning among participants.

11.4.3 Knowledge Coproduction Assessment Criteria

As with any management strategies or approaches, it is necessary to evaluate the effectiveness of the knowledge coproduction nexus partnership. Henrick et al. (2017), within the context of educational research and practice, proposed a five-dimension framework to evaluate the effectiveness of research–practice partnerships: (i) building trust and cultivating partnerships; (ii) conducting rigorous research to inform action; (iii) supporting the partner practice organizations in achieving its goals; (iv) producing knowledge that can inform improvement efforts more broadly; and (v) building the capacity of participating researchers, practitioners, practice organizations, and research organizations to engage in partnership work (Table 11.2).

Table 11.2 Assessment criteria for research–practice partnerships (Adapted from Henrick et al. 2017).

Dimension	Indicators
Building trust and cultivating partnership relationships	• Researchers and practitioners routinely work together. • Establish routines that promote collaborative decision making and guard against power imbalances. • Establish norms of interaction that support collaborative decision making and equitable participation in all phases of the work. • Recognize and respect one another's perspectives and diverse forms of expertise. • Partnership goals take into account team members' work demands and roles in their respective organizations.
Conducting rigorous research to inform action	• The research addresses problems of practice facing the practice organization. • Establish systematic processes for collecting, organising, analysing, and synthesizing data. • Decisions about research methods and designs balance rigor and feasibility. • Clarify and specify problems of practice prior to identifying and assessing strategies for addressing those problems. • Findings are shared and meet the needs of the practice organization.
Supporting the partner practice organization in achieving its goals	• Provide evidence to support improvements in the partner organization. • Help the practice organization identify productive strategies. • Inform the practice organization's implementation and ongoing adjustments of improvement strategies.
Producing knowledge that can inform improvement more broadly	• Develop and share knowledge and theory that contributes to the research base. • Develop and share new tools that can be adapted to support improvement work. • Develop dissemination plans that support partnership goals and for broader communities.

(Continued)

Table 11.2 (Continued)

Dimension	Indicators
Building the capacity of participating researchers, practitioners, practice organisations, and research organisations to engage in partnership work	• Develop professional identities that value engaging in sustained collaborative inquiry with one another to address persistent problems of practice. • Assume new roles and develop the capacity to conduct partnership activities. • Participating research and practice organisations provide capacity building opportunities to team members. • Contribute to a change in the practice organization's norms, culture, and routines around the use of the knowledge coproduced. • Allocate resources to support partnership work. • Establish conditions in the practice organization that lead to sustained impact beyond the life of the partnership.

Norström et al. (2020) stressed that the evaluation of knowledge coproduction should be socially relevant and solution oriented, sustainable and future scanning, diverse and deliberative, reflexive and responsive, rigorous and robust, creative and elegant, honest and accountable. They proposed a set of criteria for assessing the knowledge coproduced, as summarized in Table 11.3.

In comparison of Tables 11.2 and 11.3, while Henrick et al. (2017)'s assessment criteria are processes oriented, Norström et al. (2020)'s assessment criteria are built around the four principles of knowledge coproduction. Both sets of assessment criteria are equally good and in fact the points of emphasis are similar. Evaluation is performed while coproduction is in progress.

Table 11.3 Knowledge coproduction assessment criteria.

Principle	Assessment criteria
Context-based	1) A coproduction process is effectively situated within a particular place, set of relationships or a particular issue. 2) The knowledge coproduction originates from an actor already encountering the problem addressed. 3) The coproduction process is utilizing, building upon, and strengthening existing skills and relationships between participants already working in the context. 4) Outputs developed still being implemented by the participants after the project is finished.
Pluralistic	1) quantitative indicators: involvement of actors across multiple axes, degree of trust built. 2) Use qualitative indicators: written reflections and blogs; short, periodic surveys to evaluate the group dynamics.
Goal-oriented	1) Use short-term, medium-term, and long-term indicators to measure achievement of goals at individual and organizational levels. Such as enhanced capacity and attention, application of the knowledge coproduced, change of policy, change of culture. 2) Use quantitative and qualitative indicators.

Table 11.3 (Continued)

Principle	Assessment criteria
Interactive	1) Assess the frequencies and timing of encounters. 2) Assess level of participations by interviews and surveys; 3) Assess learning and application of knowledge coproduced by using quantitative or qualitative indicators.

11.4.4 Knowledge Coproduction Mindful Points

There are several mindful points of knowledge coproduction. The first is the increased transaction costs due to complexity and number of participants. This can be overcome by using techniques such as stakeholders mapping and network analysing as well as applying a step-wise approach. The second point is power dynamics in the participator processes, which can be dealt with by using some tools like Power Cube to help participants map the different ways power manifests itself. The third point is that the context or goals may be evolved or changed along the progression of the knowledge coproducing process. This may be overcome by constantly reflecting and reviewing the process to unearthing the vision, understanding the value of the participants, and developing ameliorative strategies (Norström et al. 2020).

11.5 Case Studies

Three cases are chosen as examples (Table 11.4). Case 1 is government–industry–university joint research project on improving public building energy performance. Case 2 is a university–industry–government joint research project on public building energy retrofitting. Case 3 is a university–government joint research project on building sustainability index. The three cases are analysed from eight issues: real-world problem definition, scientific research

Table 11.4 Description of the three knowledge coproduction cases.

No.	Issues	Case 1	Case 2	Case 3
1	Real-world problem definition	The actual energy consumption in public buildings is 2–3 times of the design intend	Lack of capital funding to retrofit public office buildings. Lack of project management and quality assurance strategies for building energy retrofitting.	Unsure if a mandatory green building development policy has been effective.
2	Scientific research problem description	What are the causes leading to the above-mentioned problems? How can the energy performance gap between design intent and reality use be reduced?	What are the different funding mechanisms and how can they be used to fund public building energy retrofit? What are the barriers and management strategies for this particular type of project?	How to assess the efficacy of mandatory green building development policy?

(Continued)

Table 11.4 (Continued)

No.	Issues	Case 1	Case 2	Case 3
3	Research aim and objectives	To develop strategies to close energy performance gaps from project life cycle and stakeholder perspectives: (1) understand the causes; (2) understanding how design, construction, and operation contribute to the energy performance gap; (3) develop strategies to close the gap.	To overcome the barriers in public building energy retrofitting: (1) identify the barrier and overcoming strategies; (2) develop suitable funding mechanism; (3) develop quality assurance and project management strategies.	Develop methodology for assessing the efficacy. Provide strategic recommendations for the government departments to improve the policy.
4	Critical theories	Project life cycle assessment, stakeholder management; risk management.	Different financing mechanisms; project management body of knowledge; total quality management.	Balanced score card methods; stakeholder management.
5	Methods for data collection	Interview; workshop; case analysis.	Workshops with project stakeholders; Case study	Internal stakeholder interview. Policymaker and user survey. Green building occupant survey.
6	Data analysis methods	Qualitative content analysis; quantitative statistical analysis. System dynamic simulation; Interpretive structural modelling method.	Qualitative content analysis; quantitative statistical analysis; system dynamic simulation.	Qualitative content analysis; quantitative, statistical analysis.
7	Knowledge coproduction	Workshops with project partners and funding bodies: process, methods, progress reporting, findings and reflections. Jointly writing and publishing research outcomes with project partners and funding bodies.	Workshops with project partners and funding bodies: process, methods, progress reporting, findings and reflections. Jointly writing and publishing research outcomes with project partners and funding bodies.	Workshops with project partners and funding bodies: process, methods, progress reporting, findings and reflections. Jointly writing and publishing research outcomes with project partners and funding bodies.
8	Knowledge and theory dissemination	Recommendations trialled by project partners and funding bodies. Report and recommendations launched on project partners' public websites. Papers published.	Recommendations trialled by project partners and funding bodies. Report and recommendations launched on project partners' public website. Papers published.	Recommendations trialled by project partners. Report and recommendations launched on project partners' public website. Papers published.

problem description, research aim and objectives, critical theory, data collection methods, data analysis methods, knowledge coproduction, knowledge and theory dissemination.

11.6 Methodological Framework

Based on the above-presented literature review and case studies results, a methodological framework is developed, as shown in Figure 11.2. The framework consists of seven aspects.

1) Discover and define problems from real-world practice with industry partners.
2) Convert the practical problems into scientific research problems that are based on critical theories and clearly articulate research aims and objectives.
3) Decide on research methods for data collection and data analysis.
4) Collect and analyse data and coproduce knowledge in partnership with industry and wider community.
5) Present, discuss, and disseminate research results with industry partners and the wider community through workshops, seminars, and educational champions.
6) Develop guidelines and standards and apply for awards.
7) Define new research problem and develop new research project.

The following sections provide detailed explanation to each aspect and demonstrate how they differ from 'traditional research'.

1) **Discovering the practical problems.** Properly designed workshops and discussion sessions need to be held with industry partners to discover and define the practical problems that the industry partners are facing. When defining practical problems and specific outcomes, researchers need to take the time factor into account. The industry partners normally require quick answers to their problems; they may not worry too much about theories but more about solutions to their problems. They normally want

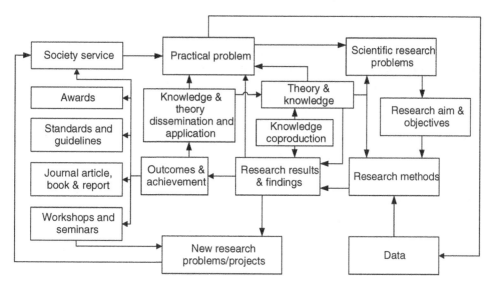

Figure 11.2 Methodological framework for research–practice nexus and knowledge coproduction.

answers and applications within 1–2 years. It is important to establish direct communication with practitioners at the right level who may have strategic and systematic thinking and decision-making capability.

2) **Defining the scientific research problems**. It is not easy to convert the practical problems into significant research problems. To do so, there is a need to understand the real-world problem and relevant theories. This means to understand what existing theories can be used to explain the real-world problem, what are the shortcomings or inadequacies of these theories, and what else is needed.

3) **Articulating the research aims and objectives.** As for any research, it is essential to clearly define the aim of the research, and based on the overall aim are the detailed research objectives. The aim and objectives provide guides and measurements to keep the research directions and progress in alignment with what is needed by the industry partners and for theoretical development.

4) **Reasoning the research methods**. More often than not, one may simply pick a method and carry on with the research, but it is necessary to provide information why certain methods are selected over the other ones.

5) **Coproducing new knowledge and theories**. The main differences between research–practice nexus research and the traditional research process are coproducing knowledge and theories, which are directly validated by immediate application in real-world settings. At the same time, the knowledge, pattern, and intellectual property (IP) developed through the partnership project should be jointly owned by the research team and the industry partners. The processes and evaluation of knowledge coproduction can be referred to the relevant contents in the previous sections.

6) **Applying coproduced knowledge and theories to practice**. It is only through practice that the theories and knowledge can be tested and proved, real impact and effect realized, and shortcomings of the theories identified and rectified. There are three issues in dissemination and application of new theories to practice. The first is validation because it is expensive to verify theories and models. The second is lack of practitioners' trust because most of the validation process did not involve practitioners. The third is that methods or models are either overcomplicated or oversimplified. To solve this problem, the researchers could regularly listen to the practitioners' advice and modify the theory and model; at the same time, practitioners could regularly provide industrial data for the research.

7) **Evaluating the short-term and long-term effects.** It is through practice that the short-term and long-term impacts and effects of the theories and knowledge can be evaluated. The short-term effect is normally reflected by change of behaviour, change of process, or change of management requirements. The long-term effect normally refers to change of police or change of culture. Both short-term and long-term effects may include economic impact, environmental impact, and social and human resource capability impact.

11.7 Summary

This chapter is perhaps one of the unique chapters in this book. The fundamental rationale of this chapter is to enhance the nexus between research and practice and to coproduce knowledge that would be more applicable and effective in solving real world problems through a research–practice nexus. As such, the chapter first discussed the gap that existed between research and practice and addressed the need to close the gap. This was followed by presenting five principles for assessing the research–practice nexus: contribution of research to practice, theoretical approaches, methodology, data analysis and synthesis, and self-conscious integration of theory and practice. The chapter then focused on knowledge coproduction, from basic concept and scope to principles and processes, as well as assessment criteria and some mindful points. First-hand case study was presented to demonstrate the key points and questions. We concluded by presenting a methodological framework that is reasonably complicated and complex but shows the details of research–practice and knowledge coproduction as a loop: society service – practical problem – scientific research problems – research aims and objectives – theory, knowledge, and knowledge coproduction – data – research results and findings, knowledge, and theory dissemination – outcomes and achievement – awards, publications, seminars, and new research problems which also close the loop. Each of these key points has been discussed in detail in the chapter. We suggest that readers pay particular attention to identifying and clarifying practical problems and knowledge coproduction processes and assessment criteria.

Review Questions and Exercises

1 What does the term research–practice nexus mean?
2 How can a research–practice nexus be achieved?
3 What is the effect of a research–practice nexus?
4 What is meant by knowledge coproduction?
5 How are research–practice nexus projects implemented?
6 How do you evaluate the effectiveness of research–practice nexus project outcomes?

References

Baarts, C. (2009). Collective individualism: the informal and emergent dynamics of practising safety in a high-risk work environment. *Construction Management and Economics* 27 (10): 949–957.

Berbegal-Mirabent, J., Sánchez Garcíab, J.L., and Ribeiro-Soriano, D.E. (2015). University-industry partnerships for the provision of R&D services. *Journal of Business Research* 68 (7): 1407–1413.

Boaz, A., Hanney, S., Borst, R. et al. (2018). How to engage stakeholders in research: design principles to support improvement. *Health Research Policy & Systems* 16 (1): 1–9.

Böckel, A., Nuzum, A.K., and Weissbrod, I. (2021). Blockchain for the circular economy: analysis of the research–practice gap. *Sustainable Production and Consumption* 25: 525–539.

Bolander, W., Chaker, N.N., Pappas, A., and Bradbury, D.R. (2021). Operationalizing salesperson performance with secondary data: aligning practice, scholarship, and theory. *Journal of the Academy of Marketing Science* 49 (3): 462–481.

Editorial, N. (2018). The best research is produced when researchers and communities work together. *Nature* 562 (7725): 7.

Evans, M. (2009). *New Directions in the Study of Policy Transfer.* Taylor & Francis.

Henrick, E.C., Cobb, P., Penuel, W.R. et al. (2017). *Assessing Research–Practice Partnerships: Five Dimensions of Effectiveness.* New York, NY: William T. Grant Foundation.

Joyce, K.E. and Cartwright, N. (2020). Bridging the gap between research and practice: predicting what will work locally. *American Educational Research Journal* 57 (3): 1045–1082.

Kerner, J.F. (2008). Integrating research, practice, and policy: what we see depends on where we stand. *Journal of Public Health Management and Practice* 14 (2): 193–198.

Kwon, S. (2022). Interdisciplinary knowledge integration as a unique knowledge source for technology development and the role of funding allocation. *Technological Forecasting and Social Change* 181: 121767.

Makin, S. (2021). The research–practice gap as a pragmatic knowledge boundary. *Information and Organization* 31 (2): 100334.

Moses, S., El-Hamalawi, A., and Hassan, T.M. (2008). The practicalities of transferring data between project collaboration systems used by the construction industry. *Automation in Construction* 17 (7): 824–830.

Neal, J.W., Neal, Z.P., Kornbluh, M. et al. (2015). Brokering the research–practice gap: a typology. *American Journal of Community Psychology* 56 (3): 422–435.

Norström, A.V., Cvitanovic, C., Löf, M.F. et al. (2020). Principles for knowledge co-production in sustainability research. *Nature Sustainability* 3 (3): 182–190.

Pizmony-Levy, O., McDermott, M., and Copeland, T.T. (2021). Improving ESE policy through research–practice partnerships: reflections and analysis from New York City. *Environmental Education Research* 27 (4): 595–613.

Podgorodnichcnko, N., Edgar, F., and Akmal, A. (2022). An integrative literature review of the CSR-HRM nexus: learning from research–practice gaps. *Human Resource Management Review* 32 (3): 100839.

Tsoukas, H. and Mylonopoulos, N. (2004). Introduction: knowledge construction and creation in organizations. *British Journal of Management* 15 (S1): S1–S8.

Unerman, J. and O'Dwyer, B. (2010). The relevance and utility of leading accounting research. ACCA research report 120.

Ye, S. (2021). *Talking Science to Schools: Organizing the Research–Practice Nexus in Early 21 St Century American Education.* The University of Chicago.

Zou, P.X.W. and Sunindijo, R.Y. (2015). *Strategic Safety Management in Construction and Engineering.* John Wiley.

12

Managing the Researching–Writing–Publishing Journey

12.1 Introduction

This chapter serves as an extension of Chapter 9, 'Journal Article Writing and Publishing', and a foundation for Chapter 13, 'Improving Impact and Citation of Research Outcomes'. This practical chapter discusses important aspects of developing a manuscript for submission to international journals and provides strategies for improving research impact and citation. These include criteria and steps for selecting topics and guidance for writing the abstract, introduction, literature review, and conclusion. It addresses the importance of having a theory to guide practice and improving theory by reflecting on practice in research writing and publishing. It also discusses the relationships between research methodology, methods, processes, and design, and thinking versus writing. Finally, it discusses how to generate research impact and citation of the published articles.

Good research normally has the following features: research aims clearly defined; research design thoroughly planned; research process properly detailed; ethical standards suitably applied; findings unambiguously presented; conclusions fully justified; researcher's experience appropriately reflected; theoretical contribution and practical implication completely discussed, and limitations frankly revealed. This chapter includes 13 parts:

1) Focusing on current major challenges and clearly defining research problems.
2) Writing with a purpose and target audience in mind.
3) Naming a good title.
4) Crafting a concise abstract.
5) Starting with an eye-opening introduction and ending with a justified conclusion.
6) Undertaking a comprehensive and critical literature review.
7) Applying appropriate research methodology, method, design, and process.
8) Writing and presenting results and discussions.
9) Referencing and acknowledging.
10) Thinking versus writing.
11) Early starting.
12) Publishing your writing.
13) Improving citations of your publication.

Research Methodology and Strategy: Theory and Practice, First Edition. Patrick X.W. Zou and Xiaoxiao Xu.
© 2023 John Wiley & Sons Ltd. Published 2023 by John Wiley & Sons Ltd.

12.2 Focusing on Current Major Challenges, Defining Research Problems

Good academic writing starts with a thorough understanding and focus on current major practical or theoretical challenges, and precisely defining and describing the problems for research in scientific forms. It is said that 80% of the time should be used for defining the problem and 20% for solving the problem. If the problem was defined incorrectly the solution will be of no use. To gain a thorough understanding of current key challenges we need to be familiar with the literature and have conversations with key industry leaders and managers. We need to go out there to the frontline of the current practice, think deeply about the strategic needs of the country, and talk to practitioners to understand how they think and the pain points. Most importantly we must grasp the key bottleneck problems and challenges practitioners are facing, describe these challenges and problems in great details, then turn the challenges and problems into scientific problems to be researched, and establish research aims and objectives. Figure 12.1 shows the cyclic iteration relationship between the realm of theory and the realm of practice (Zou et al. 2014).

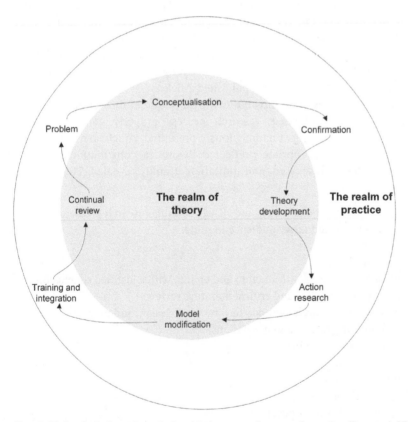

Figure 12.1 Cyclic iteration relationship between theory and practice (Zou et al. 2014 / with permission of Elsevier).

Figure 12.1 shows that the first step is to identify the practical problems with the practitioners, then conceptualize the problems in theoretical perspective and expressions, which are then confirmed with practice. Based on the confirmed research problems, theory is developed and action research implemented with practice, where modification may be carried out according to the effects of the theory on practical implementation. Following the theory modification is the training and integration of the theory with practice. Continued review is carried out for continuing improvement and discovery of new problems.

12.3 Writing with a Purpose and Target Audience

It is important to have an audience in mind and establish a clear purpose when writing a journal manuscript. During the writing it is easier and more meaningful to have an audience in mind, and think about what, why, and how to write for that particular group of audience. When we write personal letters, for example letters to a girlfriend or boyfriend or partner or parents or friends, we have a purpose of writing – they're all different people, and we frame how we write, and choose the appropriate language.

In academic writing, we are also writing for people to read and understand, and we must keep the audience in mind during the writing process. Our writing needs to make sense to the audience. This seems straightforward but has often been forgotten. In academic writing, our audience is usually peer academics or research students. It may also include professionals and practitioners. We need to think about the following questions:

- What content would they be interested in reading?
- What level of knowledge might they have relating to the context of this writing?
- How would they understand the writing?
- What are the key points of the writing?
- What are the methods used?
- How may they repeat the method if they wanted to?
- What theory or findings can be drawn that others may understand and use in their research or practice?

12.4 Choosing a Good Title

It is important to have a good title in academic writing, particularly for a journal article. When we search for literature on the Web of Science or Science Direct or Scopus, or the like, the first thing that appears on our browser is the title of a journal article or conference paper, which we read to judge if it is useful and relevant to our research. Only if the answer is yes will we go on reading the keywords and abstracts; and after that if we still feel interested in it we may download and read the full article. This process reinforces the importance of deciding on a good title when we write papers for publishing in journals. The title must reflect the contents of our writing, it may emphasize a new method, problem, perspective or finding, or simply present a major question. We may use different styles and wordings, or other more innovative ideas. It is important to have a working title at the beginning and come back to check and modify the title at the end of the writing, so that the title is punchy, reflective, and eye-catching as well as appealing.

Authors should think deeply when deciding the titles for their thesis, journal articles, or conference papers. Here is an example of a good title for journal article. The paper title is 'Understanding the key risks in construction projects in China', published in the *International Journal of Project Management*. This paper has become a highly cited paper and if researchers look at the title, they see it is about key risks, in construction and, more specifically, it is about China. At the time when the paper was published, every country was interested in the Chinese construction. In addition, there is a great deal of substance and knowledge contributions that the paper offers. It is the first paper that discussed a project's five objectives (time, cost, quality, safety, and sustainability) together and developed a three-dimensional lifecycle risk management framework that includes project stakeholders, project life cycle, and project objectives. This was innovative and integrated thinking at the time.

Generally, journal paper titles can be divided into the following five categories:

1) **Titles that emphasize the highlights of the paper**. This type of title directly shows the highlights of the research findings. For example, in the paper titled 'Not all project ambiguity is equal: A typology of project ambiguity and implications for its management', the highlight is 'Not all project ambiguity is equal'.

2) **Titles with a verbal structure**. Verbal structure can make the subject of the title more prominent and vivid. The titles of papers in many top journals often use the verbal structure, such as 'Understanding the key risks in construction projects in China', 'Monitoring precast structures during transportation using a portable sensing system', and 'Modelling stakeholder-associated risk networks in green building projects'.

3) **Titles consisting of objectives and methods**. This type of title presents the objective in the first half and then introduces the research method with 'through' or 'via', such as 'Closing the building energy performance gap through component level analysis and stakeholder collaborations', and 'Intelligent bridge management via big data knowledge engineering'.

4) **Titles with advanced instruments/tools**. The advanced instruments/tools mentioned in such titles are often very attractive to readers, such as 'Exploration of public stereotypes of supply-and-demand characteristics of recycled water infrastructure – Evidence from an event-related potential experiment in Xi'an, China', 'BIM data flow architecture with AR/VR technologies: Use cases in architecture, engineering and construction', and 'Semantic localization on BIM-generated maps using a 3D LiDAR sensor'.

5) **Titles that emphasize the research problem**. This type of title directly presents research problem. For instance, 'Why do individuals engage in collective actions against major construction projects? – An empirical analysis based on Chinese data', and 'How safety leadership works among owners, contractors, and subcontractors in construction projects'.

12.5 Crafting a Concise Abstract

The importance of a good abstract cannot be stressed enough. Statistics show that 90% of readers may only read the abstract, hoping to know the whole story of the article. The abstract can be seen as a mini version of a journal article. Moreover, abstracts are published in the public domain by the publisher, so a good abstract will attract more readers and citations. The question is what to include in an abstract? What makes a good abstract?

Readers appreciate it when we tell them what we are going to tell them, and then tell them what we have told them. A good abstract informs the readers of the importance of the research, the methods used, the main results and discussions, the recommendations and conclusions, and the implications on contributions to knowledge, theory development, or current practice improvement. The abstract is normally six to seven sentences up to 250 words. Some journals may limit an abstract to be less than 100 words while some may allow an extended abstract to be up to 1000 words.

A graphical abstract is used to present the most important information about the research work in a clear and concise manner and is as important as the title of the paper. Most top journals, especially those with a high impact factor, require authors to provide a graphical abstract. A graphical abstract is an important part of papers. It consists of a diagram and a paragraph of text. A graphical abstract allows the authors to give a high-level overview of the research and the main innovations in the paper, attracting the reader's attention and allowing them to quickly understand the paper.

When to write an abstract? Some people suggest writing the abstract at the end, but we would suggest having a working abstract to guide writing and to bring the authors back if they are lost or wondering off in the writing process. At the end of the writing, researchers need to come back and revise the abstract to make it reflective, concise, and attractive for readers.

12.6 Starting with an Eye-opening Introduction and Ending with a Summative Conclusion

It is important to have a good introduction and good conclusion. Introducing research is like opening up a story and unfolding an important issue. It has to be interesting and draw readers' attention. All writing is about telling a good story. The question is *how*. The introduction may include a brief literature review to establish the significance, gaps, and needs of current research, which then prompts the research problem statement, and the research aims and objectives. We normally state the aim of the research at the end of the introduction.

There are differences between an introduction and an abstract. While an abstract could be seen as a mini version of an article that is self-contained and complete, an introduction takes readers into the scene, opens up the story, and establishes the significance and needs. It should ignite readers' imagination about the topic, to attract them and make them want to read the rest of the paper.

The conclusion section is where researchers can make some solid points that echo the introduction, provide answers to the research aims or summarize what has been discovered by the research, such as the new knowledge or theory developed. It is about improving the understanding of the phenomenon being researched, and how this new knowledge and theory may help improve practice. Conclusions need to be justified and concise, and should not include new information, but sometimes can be generalized to a higher level by demonstrating the significance and outcome of the research as well as implications to knowledge advancement and practice improvement. When writing the conclusion, ask yourself the following questions:

- Have the research aims and objectives been achieved?
- What are the main points of the results?

- What theory was developed or what is the theoretical contribution?
- What is the implication of these results to practice?
- What is the limitation of this research?
- What is the future research direction?

12.7 Undertaking Comprehensive and Critical Literature Review

The first part of a research is to review current literature, analyse and summarize what theories have been developed, what research methods have been used, and what broad research has been carried out. The literature review should establish research gaps and the needs for further research, which become the current research problem statements. Review should cover three types of relevant literature: empirical literature, theoretical literature, and methodological literature.

It is important to structure the literature review in a logical presentation and layout. The literature review may include the main theories and body of knowledge, key points, chronical historical development, main authors and their collaboration networks, current research focus, and future research trends and needs. It is necessary to provide evidence to support the above-mentioned points. It is important to include mainstream authors' work, particularly their recent publications.

Several software packages can be used to help analyse the literature both quantitatively and qualitatively, such as bibliometrics and scientometrics analytical software; some examples are: 'Review of 10 years research on building energy performance gap: Life-cycle and stakeholder perspectives published in *Energy and Buildings*' (Zou et al. 2018), 'A mixed methods research design for bridging the gaps between research and practice in construction safety management, published in *Safety Science*' (Zou et al. 2014), and 'A science mapping approach based review of construction safety research, published in *Safety Science*' (Jin et al. 2019), as well as 'Computer vision technologies for safety science and management in construction: A critical review and future research directions, published in *Safety Science*' (Guo et al. 2021). Figure 12.2 is a flowchart for scientific review of literature.

When reviewing the literature, it is important to be critical and inclusive, with different perspectives, and to be forward thinking, to provide future research direction with a conceptual framework, proposition, or hypothesis. Make sure that the review is comprehensive and complete, and covers most of the relevant theories.

It is possible to develop a good journal article purely based on literature review, if the review is carefully designed and implemented. The main contributions of this type of paper are summarizing previous work and the current body of knowledge, developing a new framework, and pointing out future research needs with potential research methods. Section 9.3.5 of Chapter 9 provides details on how to undertake a comprehensive and critical literature review.

12.8 Research Methodology, Method, Design, and Process

Research methodology is an essential part of developing and implementing a research project or writing articles for international peer-reviewed journals. It is important to understand the relationships between worldview, philosophical assumption, research

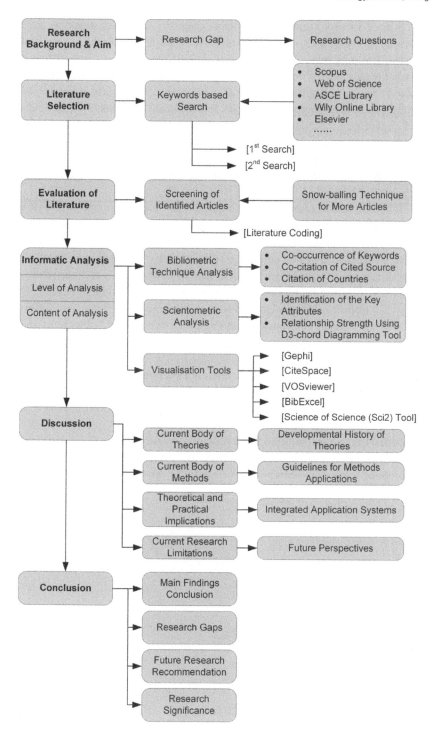

Figure 12.2 Process of conducting a literature review.

methodology, research method, and the knowledge created. To achieve research aims and answer research problems we could use quantitative or qualitative methods, or mixed qualitative–quantitative methods. Quantitative and qualitative methods have different strengths and weaknesses; mixed methods research, which combines quantitative and qualitative methods, has gained growing interest and importance in recent years. Mixed methods sometimes are called pragmatic methods, which means the selection and determination of which research method or methods is dependent on the research aims and objectives, or research questions. It gives less consideration of the worldview and philosophical assumption, but is acceptable as long as the research method adapted can solve the research problem or answer the research question (and hence achieve the research aims) essentially and sufficiently.

A complete and good research article should discuss the fundamentals of research methodology. It should justify the choice of the research methods, and discuss data collection and analysis methods. There are several important points that an article may discuss, including worldviews, ontology, epistemology, methodology, method, source; constructivism vs interpretivism; objectivism vs positivism. The hierarchical relationship between philosophy-methodology-method-content-context is shown in Figure 12.3. It is also important to understand that research methodologies are different from research methods, the relationship between research aim and objectives and research methods, and when and how to use a qualitative method, quantitative method or mixed methods.

Ultimately, the decision and reasoning for selecting a particular research method or mixed methods depend on the complexity of the research problem and the research aims and questions. Chapter 4 provides a useful mixed research methodological framework, as shown in Figure 4.1.

Another aspect that needs consideration is the research–practice nexus. To facilitate the application of research outcomes into practice, there is a need to embed research with practice and coproduce knowledge with practice. Chapter 11 provides useful frameworks, as shown in Figure 11.2.

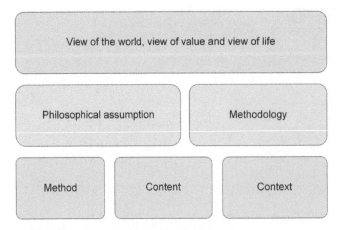

Figure 12.3 Hierarchical relationship between philosophy-methodology-method-content-context.

12.9   Presenting Research Results

It is obvious that everyone would hope to achieve good results after a long research journey. In terms of presenting the results, we could consider using tables or diagrams such as pie charts, flow charts or other graphical presentations. If the results are long, it is useful to break them into subsections, using subheadings so that the article structure is easy to follow, understand, and remember.

The most important thing, however, is to ensure the results are correct and justifiable. The best results may be within the common reasons but outside the current thinking. It is necessary and essential to discuss the results by explaining what they mean and how they relate to the current literature, if they agree, contradict or complement the current literature; or if there is a surprise new result that no one has thought about. From the results, it is useful to produce a theoretical framework, or a set of theoretical statements, which also form a theoretical contribution to the field of research.

12.10   References, Acknowledgment, and Ethics

We should not underestimate the importance of including sufficient references and using the proper format in our manuscript for submission to journals. Often reviewers look at the references first to see the number of references included, and whether the key ones have been included. Only by reading enough information in the current literature can we develop a good research project and achieve good results. Having said that, we also need balance – not too many, not too few, just sufficient.

Remember, different types of references have different credibility. Nowadays information is at our fingertips and we can easily get lots of information from the Internet; we need to justify which ones are credible, important, and useful for our research. Reference selection is a tough and challenging task. Researchers need to search widely to include as many references as possible and then filter them to select those on the same themes as the research. It is important to thoroughly read and truly understand the actual meaning of the articles when citing them. It is important to chase to the source references instead of citing other people's citing.

There are different styles to present the references at the end of the paper. If researchers have decided to submit their paper to a particular journal it is a wise idea to check the requirements, or 'authors guides', so that the writing fits in. Researchers may check if their target journal has published similar papers and, if so, try to cite the relevant articles from that journal. It is helpful to cite relevant references from the target journal. It is also definitely important to point out the 'points of differences or departure' of the paper with the already-published ones.

It is necessary to acknowledge the grant body or organization which provided funding to the research. It is also useful to thank reviewers, interviewees, survey respondents, and other important people who have helped in the research journey, but do not overkill on this part because it could make things look unreal and may damage the authenticity of the research.

Finally, ethics has become ever more important. There are requirements about ethics, which means that researchers need to declare any conflict of interest. Discussions on research ethics have been addressed in Chapter 1 and other chapters. It is a requirement to keep the source information and key calculation processes in case readers want to replicate the research in the future. This is a new requirement applicable for all journals now.

12.11 Thinking versus Writing

Writing is, in fact, a reflection of our thinking. There is critical thinking, creative thinking, and reflective thinking; there is also logical thinking, forward thinking, and balanced thinking. Some of these concepts are explained here.

- Critical thinking is to critique our own work or others' work. Questions like 'Does it make sense?' or 'Can it be done differently?' are often asked in our head when reviewing previous literature. This means we need to think critically, not simply accept things at face value.
- Creative thinking allows us to come up with something new, either a new method, new concept, new product, new idea, or new process.
- Reflective thinking is to reflect on what we have done and find better ways for future actions.

Figure 12.4 provides an overall picture of these thinking styles. Depending on the research problem and context, we may use one or more of these thinking methods, and ask five 'why' questions to find the root causes and answers, which may lead to new questions.

We also need to develop our logical thinking, forward thinking, and balanced thinking. Everything needs to be reasoned and logical. We need to look and think forwardly. It is important to balance our thinking. Ultimately, we should develop our system thinking capability, in other words, think from a systematic perspective. Figure 12.5 shows the

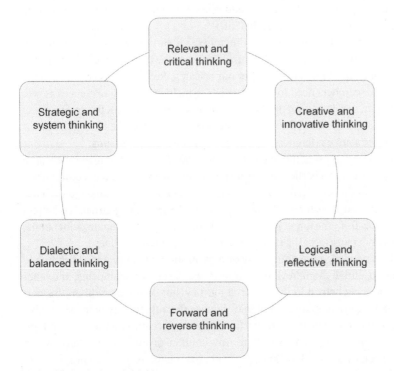

Figure 12.4 Styles of thinking.

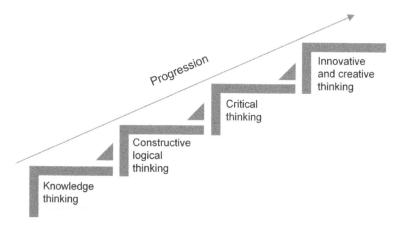

Figure 12.5 Thinking development roadmap.

thinking progression levels: from knowledge thinking and constructive logical thinking to critical thinking and, ultimately, innovative and creative thinking.

12.12 Early Start Is Half Done, Only Done Is Done

It seems common for common people to take a long time to start something, as part of human nature. We tend to be lazy and therefore it may take a long time to start writing a paper, doing an assignment, fixing the roof, tidying up the garden, etc. Finding motivation and methods to actually start early is important. Early start is half done! People normally say a good start is half done, but an early start is equally important as half done, because once researchers start they can revise and improve, and have time to do so.

From a time-management perspective, it is a good idea to start as early as possible. When we plan time, we need to keep some lead time together with wrap up and revision time to ensure the work that we have done is of the best quality. We need to plan for the time to complete the first draft, cool down, revise the draft, then submit.

We need to remember the rule of thumbs of 'now or never'. If we do not start right at the time when we have the idea, other things may get into the way and we may not have time to return to this writing. It is important to start while the ideas are still fresh in mind, the 'hot stove' principle.

We would suggest to first write down ideas and then do some literature review; use a table to summarize what has been discussed in the literature. Then, move on to define the research needs or problem statements, form the research aims that can be turned into research questions. From there, we can effectively implement and make steady progress in the writing. It is helpful to have a time plan for writing, treating the writing as a project, and apply project schedule and management skills to the writing. We would also suggest that we set milestones and deadlines and stick with them. When collaborating with other researchers, it is helpful to collaborate with someone who has complementary skills and is a good team player, so that researchers can support each other and provide peer pressure to move forward.

Again, it is always easier said than done. Academic research writing is just like learning how to swim – researchers can easily learn about the theory, but unless they actually jump into the swimming pool and practice swimming they may not even be able to keep themselves floating on the water. Researchers might need to sink in the water, have a few water-chokings, practice, and more practice, before they can float on the water. They may swim slowly at the beginning, then improve, then becomes a swimming champion. The same goes for academic research writing.

12.13   Publishing the Writing

There are many ways of publishing writings nowadays but the most useful ones still lie in publishing with the prestige peer-reviewed international journals, particularly those with high impact factor and in Q1 or Q2 quartiles. Apart from their own quality and reputation, articles published in these journals are used as a key item and criteria for a university's international rankings. Publishing in high-quality peer-reviewed international journals has also become a requirement for PhD or Masters students, and for academic promotion. Publishing in high quality journals also helps with career development and reputation. For the very least, publishing in high-quality journals contributes to the advancement of theory and society. Publishing also increases readership of your research.

Every journal has its own guidelines and requirements for authors. It is important to become familiar with and strictly follow these guidelines and requirements when preparing and submitting manuscripts. As mentioned previously, it is useful to read and cite some relevant articles published in the journals in which you plan to submit the manuscript to. This allows researchers to get a sense of the quality, types, and styles of the articles this journal may publish whilst keeping them up to date with the current literature. In addition, it also provides a good impression to the editors by having cited relevant articles from their own journals.

Generally speaking, in addition to the manuscript, researchers also need to write a covering letter to the editor, include several highlights of their article, and suggested reviewers. In the online submission system, there will be some questions about authorship, ethics, and funding body that need to be answered.

A manuscript will normally be sent to up to five people for review, and two or three will return their comments. Their comments could be very long or short, critical or constructive, positive or negative, helpful or otherwise. In any case, researchers need to treat and respond to these comments carefully, seriously, and positively.

Once review comments are returned, one of the five outcomes may be applicable to the manuscript: accept, accept with minor revision, major revision and resubmit, reject, or suggest transfer to another journal. Researchers need to read these comments a few times and provide a comprehensive, detailed, robust, and objective response to each of the comments. Take this process as a second loop of learning and opportunity for improving the manuscript's quality and thinking. Authors are required to provide responses to each of the reviewers' comments. We suggest the response be itemized with one comment one response. Researchers may use a two-column table to respond to the reviewer comments. The first column lists the comments, and the second column provides responses, and explains what action has been taken. Sometimes the editor may also have specific comments that require a response. This must also be done accordingly.

In the case that two or more reviewers raise similar issues, it is useful to cross-check and combine them. If two or more of the reviewers have contradicting comments, researchers need to think more carefully about how to respond. Remember writing a response to the reviewers' comments is partly 'science' and partly 'art'. Providing an evidence-based objective, friendly, persuasive, and convincing response will increase the chance of the manuscript being accepted. We would suggest not to overkill by saying too many 'thank you', instead just be polite, reasonable, detailed and, most importantly, 'to the point'. Everyone is very busy nowadays, so 'to the point' will save everyone's time.

12.14   Increasing Impact and Citations of Publication

Once the article is published it is important to think about how to improve its citations and generate impact so that others may read and build upon your work. There are many ways of doing so. Some authors provide a one-line statement in their email signature block, informing the e-mail recipients about the newly published article, others use social media and networking platforms, such as Facebook, WeChat, and Twitter to self-promote. There are also ResearchGate, Google Scholar webpages, etc. There are some free research database systems that allow authors to add their articles. Some authors do self-citation but we would suggest being cautious about this. Yet other ways to improve impact and citation are to attend and present research findings and published journal articles in conferences, and communicate with practitioners.

12.15   Summary

In summary, we should remember and practice the following points:

- Focus on current major challenges (major contradicting points) and clearly define the major problems and research aims and objectives.
- Write with a purpose and have target audience in mind.
- Develop a reflective and eye-catching title.
- Craft a concise abstract; Start with an eye-opening introduction and end with a justified convincing conclusion.
- Conduct a critical review of the literature.
- Provide reasons for the choices of research methodology, methods, design, and process.
- Present results and discussions in an objective and logical way. Include sufficient references and provide an acknowledgement.
- Develop thinking and start early, and remember that only done is done.
- Remember that undertaking academic and scientific research writing is partly 'science' and partly 'art', and constant 'writing, writing, writing' is the key to success.

Review Questions and Exercises

1 How to discover and define a research problem?
2 How to ensure the logics between the introduction, research aims and data collection and analysis and results and discussions?
3 What are the different thinking styles?
4 How may different thinking styles influence writing?
5 How to write a review article and what are the structures and components?
6 How to respond to the editor's and reviewers' review comments?

References

Guo, B.H.W., Zou, Y., Fang, Y. et al. (2021). Computer vision technologies for safety science and management in construction: a critical review and future research directions. *Safety Science* 135: 105130.

Jin, R., Zou, P.X.W., Piroozfar, P. et al. (2019). A science mapping approach based review of construction safety research. *Safety Science* 113: 285–297.

Zou, P.X.W., Sunindijo, R.Y., and Dainty, A.R.J. (2014). A mixed methods research design for bridging the gap between research and practice in construction safety. *Safety Science* 70: 316–326.

Zou, P.X.W., Xu, X.X., Sanjayan, J., and Wang, J.Y. (2018). Review of 10 years research on building energy performance gap: life-cycle and stakeholder perspectives. *Energy and Buildings* 178: 165–181.

13

Improving Impact and Citation of Research Outcomes

13.1 Introduction

This chapter provides an integrated framework for improving impact and citation of research outcomes. It covers why we might conduct research and in what contexts, followed by how to prepare for publication, how to start writing, how to respond to reviewers' comments, and how to disseminate, promote, and monitor research publication. It also provides suggestions for increasing research impact and citation as well as how to collaborate in research internally and externally, locally, and internationally. In short, it is about how to improve quality, productivity, citation, and impact of research from a longitudinal perspective as shown in Figure 13.1.

Figure 13.1 shows five aspects for conducting research from a research life-cycle perspective. These are the why, the what, the how [1–5], the collaboration, and the useful tips and considerations. The details of each of the five aspects are graphically presented in the following series of figures, in different sections together with textual explanations. Some of these figures are explained with texts when the concept is complicated; however, most of the figures are straightforward with no need of text explanation. Readers are encouraged to consider the points listed in the figures.

13.2 Rationale for Conducting Research

There are numerous positive points for *why* we should conduct research, as shown in Figure 13.2, as well potential reasons not to do research, which are also listed. In essence, in today's higher education system, conducting research is not an optional but compulsory duty of every member of the academic staff and higher degree research students. Universities use research performance as a key criterion for measuring everyone's annual performance review and promotion. It also provides personal satisfaction and sense of achievement, by developing new theory and contributing to the knowledge base. As for research students, it is part of the educational training process to write and publish research results, and preparation for a potential academic career.

However, if an academic staff member does not want to conduct research, it is also possible to find reasons not to do so, such as no time, teaching comes first, do not know how

Research Methodology and Strategy: Theory and Practice, First Edition. Patrick X.W. Zou and Xiaoxiao Xu.
© 2023 John Wiley & Sons Ltd. Published 2023 by John Wiley & Sons Ltd.

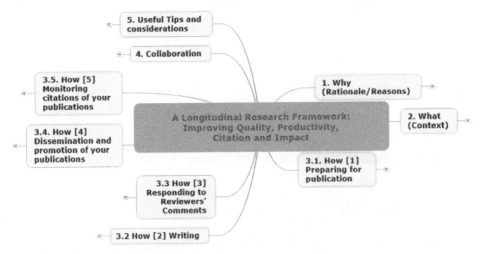

Figure 13.1 Longitudinal research framework for improving research.

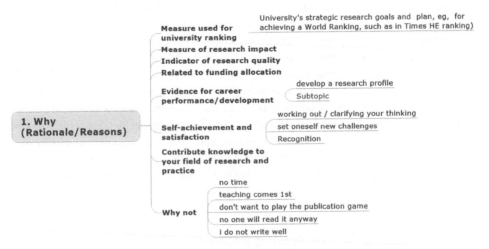

Figure 13.2 Longitudinal research framework (Part 1: The Why).

to write, or do not want to play the publication game, etc. But all these can be overcome and should be overcome with practice and time management. No one knows how to write at birth, all writers learned how to write and constantly practice to improve writing.

13.3 Research Context

The second important point is *what* to research. In the previous chapters we have discussed the importance of choosing a mainstream research topic that will bring benefit the public, the common people, and the entire society. Within this broad coverage, it is important to define and describe the objective of research. We normally start by understanding the

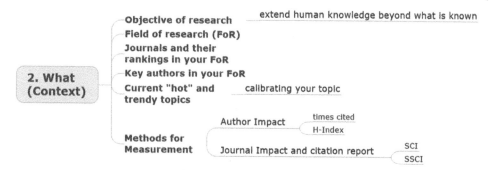

Figure 13.3 Longitudinal research framework (Part 2: The What).

broad field of the research that we are willing to spend our time on and in which we are interested in undertaking an in-depth investigation. We then start collecting and reading the relevant journal articles and conference papers that publish the research outcomes of the particular research field we have chosen (Figure 13.3).

Figure 13.3 lists some strategies for choosing journal articles, for example the journal rankings, the key authors in the research field, and the current hot trendy topics. A tip is to read some good recent review articles in the field to get some good ideas about the state and level of the research. When identifying the key authors in the research field it is useful to check their websites and profiles on public websites such as Google scholar or ResearchGate etc. These publicly open websites give pretty good measures of the international standing of an individual's research performance.

13.4 Preparing for Publication

Preparing for publication is a complicated and complex process, as shown in Figure 13.4, including choosing and focusing on an important research topic, choosing the right journal, type and contents of the research, etc. Most of these points have been discussed in the previous chapters. Readers are encouraged to use the points listed in Figure 13.4 as a checklist, and to follow through.

Choosing the right journal for submission perhaps deserves some discussion. While some people suggest submitting manuscripts to the top journals in the field, others suggest submitting to the most suitable ones. Both have good reasons. How to choose the most suitable journals depends on a number of criteria and these criteria are different for different authors. For junior researchers or research students, meeting the annual performance or completion requirements is a minimum. Only after these requirements are met, one might consider 'better' or higher-ranked journals. For senior researchers who focus on research quality rather than quantity, it is worth paying attention to the better ranked journals. Having said this, it is still challenging to know which journals have lower rejection rates (which mean higher acceptance rates). The most important criterion is always to ensure that the research is of good quality and within the scope of the targeted journals.

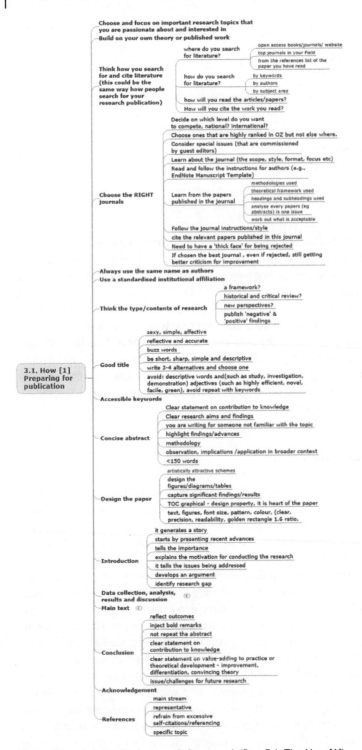

Figure 13.4 Longitudinal research framework (Part 3.1: The How [1]).

Figure 13.4 shows the typical structure of a manuscript for submission to a research journal. Several previous chapters have discussed the structure of a research article. As always, researchers need to be mindful of research ethics, which is a very important issue in today's research publication.

13.5 Process of Writing

It is important to develop a writing plan that sets a writing goal and locates time for the writing. It is suggested to start writing four levels of headings to develop a paper structure. There are many tips and strategies for writing itself, including free writing, generative writing, streamline writing, iterative writing, get-on writing, joining a writing group, and becoming a regular writer, etc. Allocating sufficient and regular time for writing is a major challenge these days as everyone is very busy with many conflicting priorities. Therefore, time commitment is a precondition for writing.

After the writing is complete, it is necessary to proofread the manuscript, either yourself or with assistance from someone else. The importance of proofreading is simple; instead of letting the journal editors or reviewers point out the mistakes in the manuscript, it is much better to let co-authors or friends point these out and have improvements and corrections made prior to the submission. Strategies for proofreading are listed in Figure 13.5.

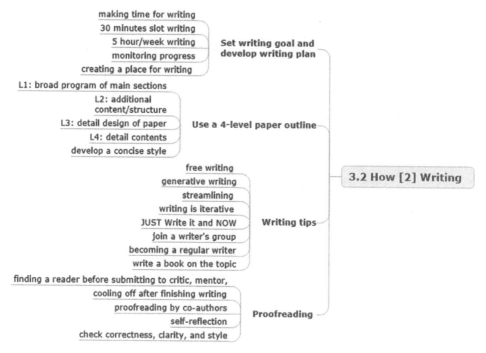

Figure 13.5 Longitudinal research framework (Part 3.2: The How [2]).

13.6 Responding to Reviewers' Comments

Manuscripts submitted to a journal will normally be reviewed by three to five reviewers, as a peer-review process. Reviewers are chosen from the journal editorial board and other sources such as 'snowball' rolling nominations, or by research keyword matching. Some journals also allow authors to suggest potential or possible reviewers. Authors should use these opportunities by providing a list of potential reviewers who would be interested in the research field and reasonably familiar with the research. It would be helpful to suggest reviewers whose work you are also familiar with and might have cited their recent publications.

Once reviewers' comments are complete, authors will receive the comments in full set, together with editors' comments and the journal's common notices. A tentative decision is normally provided about the manuscript, as one of the five possibilities: accept, accept with minor revision, major revision, resubmission, or reject. In most cases, the authors will need to revise their manuscript and respond to the reviewers' comments.

Responding to reviewers' comments is partly science and partly art. Details have been discussed in the previous chapter while Figure 13.6 provides a summative graphical representation. We suggest reading the comments carefully a few times; it may be necessary to 'cool off' after reading the comments particularly when the comments are negative or very critical. If contradicting comments are found between the reviewers' comments, researchers may write to the editor to ask which comments should be considered when revising the manuscript.

When responding to the comments, it is important to keep the authenticity of the research, which means sticking with the research aims, objectives, methods, process, and results that researchers have carefully designed and followed through. Having said so, it is important to ensure the research is rigorous and complete. At the same time, it is important to explain the research to the reviewers in a nice straightforward and polite manner. In many cases the reviewers' comments could be incorporated into the discussion section and

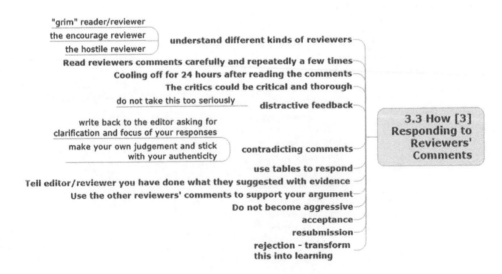

Figure 13.6 Longitudinal research framework (Part 3.3: The How [3]).

future research section of the manuscript. In the responses it is important to provide actual evidence of revision.

A key issue is how to write such a response letter, what type of language and tone to use. This is perhaps the art of responding. We are all reasonable people and using common, peaceful, fair, and equal tones is sufficient. One point to remember is not to get into an argument with the reviewers. There is no gain and hence no point for anyone to get into writing arguments. Chapters 9 and 12 also provided useful information on this topic, from different perspectives, and we encourage readers to read these chapters together.

13.7 Strategies for Disseminating and Promoting Research Outcomes

After the publication of the research outcomes and outputs, such as publication of articles in a reputable international journal, it is important to disseminate the research outcomes as widely as possible. Figure 13.7 provides some strategies for doing so, including allowing for maximum flexible re-use of the work, discussing and disseminating research publications with policymakers and practice, presenting research outcomes and papers at conferences and seminars, and featuring research results in newspapers and broadcast on social network media.

Figure 13.7 Longitudinal research framework (Part 3.4: The How [4]).

There are so many social media platforms for networking nowadays which can be used for research outcomes and publication dissemination, as listed in Figure 13.7. Choosing several of the most effective social network media in the field for dissemination would be a good strategy too. Apart from the list in Figure 13.7, some people also use WeChat, QQ, Blog, etc.

13.8 Monitoring Citation of Research Publication

Once published it is important to monitor the citations of the research publication. Figure 13.8 provides a number of methods for doing so. These include using a number of publication databases such as Wiley, Scopus, ScienceDirect, Google Scholar, etc.

Reflecting on why some of the publications are cited more frequently than others is a useful way for continuous improvement and development. One point to keep in mind is not to over self-cite. In other words, unless it is suitable, it is not a good idea to cite your own work for the sake of citing. Only when the research aims and contents are matching should researchers cite their own work; this is reasonable and good practice.

13.9 Collaboration with Peer Researchers

Collaboration with peers either internally or externally is extremely helpful for idea development, writing progress, and citation and impact improvement. Figure 13.9 provides the reasons and methods. It is also useful to develop collaborations with industry partners, particularly when the research is of an applied nature. Chapter 11 provides details on improving the research–practice nexus and knowledge coproduction.

When finding collaborators, a key consideration is complementary skills between each other, and the commitment to the research and writing. Good team members also provide good research ideas and good support to each other. Good teamwork is definitely helpful and effective in producing high-quality research and publications, but if the team is not united it could also lead to dysfunction, which would delay progress and outcomes. The worst-case scenario would be a dispute in IP and authorship and so on; this should be avoided at the beginning of team collaboration.

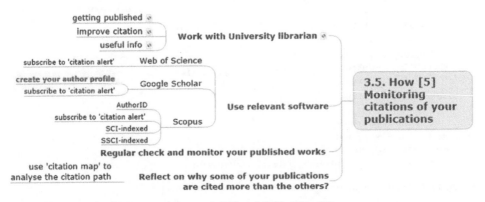

Figure 13.8 Longitudinal research framework (Part 3.5: The How [5]).

Figure 13.9 Longitudinal research framework (Part 4: Collaboration).

13.10 Some Useful Tips

Figure 13.10 provides some additional useful tips on writing for publication in high-quality journals. Again, it is all about locating sufficient time for writing, having co-authors who are familiar with the research to provide critiques and using Endnote or other software to manage references. Researchers need to be consistent and resilient in the research and writing process. There is no flat road ahead, there will be ups and downs.

The last point is about not getting too attached to what you are writing. Sometimes researchers need to sit back and think from different perspectives to provide well-rounded research outcomes, and be willing to cut unhelpful material.

Regarding open access journals, there are different views and responses to these new phenomena. Some accept and participate while others are not so keen to join this new wave and new way of publishing, for various reasons depending on individuals' actual situations and circumstances.

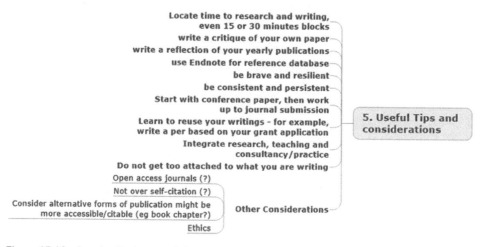

Figure 13.10 Longitudinal research framework (Part 5: Tips and considerations).

13.11 Summary

This chapter has adapted a life-cycle longitudinal perspective to describe research, with the aim of improving impact and citations of research outcomes. We started by discussing the rationale and reasons for undertaking research (i.e., the why), then moved forward to discussing the strategies on preparing for publication (The How [1]), writing (The How [2]), responding to reviewers' comments (The How [3]), dissemination and promotion of research outcomes (The How [4]), and monitoring citations of publications (The How [5]). To achieve high level impact and citations of research outcomes, the chapter also discussed the importance of strategies to collaborate with peers internally and externally, locally, and globally. The chapter finally provided several useful tips.

The rationale and reasons include positive and negative points, i.e., why and why not. Figure 13.4 is a comprehensive framework for preparing for publications that is worthy of in-depth study. Regarding writing, there are different styles and strategies; the key is to get on with it and constantly write, with a framework and structure. This framework can include three or four levels of headings and subheadings to guide your thinking and writing, so that you can easily write with a direction to guide you. Chapter 9 also provides details on journal article writing.

Responding to the editor's and reviewers' comments is partly science and partly art, which means require careful thinking, not only technically but also strategically. Figure 13.6 provides some useful strategies. Chapter 9, 'Journal article writing and publishing', also provides useful techniques, as does Chapter 12. These contents should be read together to develop your comprehensive understanding and your own strategies.

The material presented in Sections 7, 8, and 9 is particularly useful for improving research impact and citations. Such concerns are not commonly shown in research methods texts, but we believe that these points are becoming more and more important in the research community and worth discussion. These include strategies for disseminating and promoting research outcomes and monitoring the impact and citations, as well as collaboration. Collaboration is effective not only for improving impact of research outcomes but also for the research itself, as collaboration stimulates different minds and provides peer support and pressure, which leads to better research outcomes and personal development. This chapter provided some useful tips on how to find and selecting research collaborators. We hope the chapter will prove useful to many academic research careers.

Review Questions and Exercises

1 What is the rationale for conducting research?
2 What are the processes of writing for publishing in journals?
3 How to respond to reviewers' comments?
4 How to improve the impact of the research?
5 How to improve the citation of the published work?
6 How to collaborate with peer researchers?

14

Concluding Remarks and the Ways Forward

This book has attempted to cover many aspects of research methodologies and strategies. We have tried to balance depth with considered breadth. As a result, some chapters have provided more in-depth discussions while others have focused on broad coverage. We will leave this to the reader to decide if they want to continue to study the same topic area using other sources.

This book began with fundamentals of research, moving on to qualitative and quantitative research as well as mixed methods research and case study research. These are traditional methods. Based on these, the book discussed technology-enabled and data-driven research. These are new emerging research methods. We encourage readers to continue following these new topic areas because these are the future directions and trends.

Apart from these research methodologies, we discussed several strategies that will help readers improve their research productivity, quality, and impact. Topics covered include writing for publishing in international journals, writing thesis, improving research and practice nexus. We conceptualized research from a longitudinal perspective, which means viewing researching–writing–publishing as a whole journey. Finally, the book addressed the importance of improving impacts and citations of research outcomes, a final step of research.

There is no shortcut in research and writing, only hard work, breakthroughs, progression, passion, and enjoyment. Before concluding the book, we provide the following 12 critical success factors in undertaking scientific research.

1) Give great importance and priority to research.
2) Allocate sufficient time for research and manage time wisely.
3) Overcome, instead of by-passing, difficulties and bottlenecks to achieve progress and innovation.
4) Plan the content and quantity of research reading and writing every day, and be persistent and diligent.
5) Conduct regular self-reflection, by asking yourself whether improvements can be made to the research design and process.
6) Strengthen communication, discussion, and collaboration with like-minded people from different disciplines, to generate new thoughts, and expand research perspectives and thinking modes.

Research Methodology and Strategy: Theory and Practice, First Edition. Patrick X.W. Zou and Xiaoxiao Xu.
© 2023 John Wiley & Sons Ltd. Published 2023 by John Wiley & Sons Ltd.

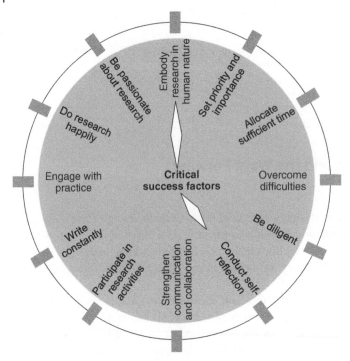

Figure 14.1 The clock model of critical success factors in research.

7) Participate actively in research activities, including seminars, conferences, and meetings, online and offline, to broaden social and research horizons.
8) Write papers and articles, obtain peer review comments, submit and publish papers and articles in journals and conferences to get the most out of your research.
9) Engage with practice, striving to integrate theory with practice and guide practice.
10) Do research happily and enjoy the research journey.
11) Research topics that you are passionate about.
12) Make research and curiosity part of your human nature.

The above 12 research critical success factors can be integrated into a clock model, as shown in Figure 14.1.

We understand that there might be limitations in the book and we encourage readers to build on this book and continue the research journey. Again, research should be a happy and continuing journey. To this end, we wish every reader happy researching.

'The way ahead is not ending, I search with my will unbending' (Chinese proverb).

Index

a

Action research   200, 212
Agent-based modelling   7, 15, 37, 53, 54, 61, 80, 151
Artificial Intelligent (AI)   19, 32
Association   136
Attitude, skill, and knowledge (ASK)   1
Axiology   5, 16, 157

b

Behaviour sensitivity test   47
Behaviour test   45, 47
Behavioural factors   122, 124, 126
Bibexcel   172
Big data   122, 129, 130, 135, 137, 147, 148, 154, 214

c

Case selection   101, 102, 103, 104, 111, 114
Case study   5, 11, 14, 15, 22, 32, 34, 61, 92, 93, 97–101, 105–110, 113, 114, 118, 206, 209
Chi-square test   41
Citation   225, 232, 234
CiteSpace   172
Classification   136
Clustering   136
Cognitive analytics   136
Collective case study   98–100
Communication   29, 30, 235, 236
Complex system   10, 11, 43, 44, 53, 70, 71, 151
Conceptualization   3, 4, 13, 15, 200, 212

Configuration   108, 111, 112, 113
Confirmation   200, 212
Construct   3, 8, 10, 42, 85, 91, 97, 98, 105–108
Constructivism   3, 4, 5, 85, 218,
Constructivist   97,
Content analysis   5, 31–35, 92, 93, 206
Contribution of research to practice   200
Control variable   38, 39, 40, 136, 191,
Covering letter   179, 222
Creative thinking   220, 221
Critical thinking   220, 221
Criticism   34, 80, 154, 155,

d

DART-Europe   197
Data analysis and synthesis   201
Data analysis strategy   29
Data cleaning   135
Data preprocessing   135
Data science   129, 130, 132, 137, 145, 148, 149, 151, 156, 158
Data storage   106, 135
Data-driven research   9, 14, 37, 80, 138, 145, 147–149
Deduction   5, 12, 16, 154
Delphi method   25
Dependent variable   38, 39, 40, 42, 100–104, 109, 117–119, 124, 125, 152, 191
Descriptive analytics   136
Descriptive case study   98, 99, 110,
Diagnostic analytics   136

Research Methodology and Strategy: Theory and Practice, First Edition. Patrick X.W. Zou and Xiaoxiao Xu.
© 2023 John Wiley & Sons Ltd. Published 2023 by John Wiley & Sons Ltd.

e

Electroencephalography (EEG)   119–122, 124–126
Embedded design   8, 86–89, 91, 93, 94
Epistemology   4, 12, 16, 157, 218
Ethnography   5, 21, 34,
Evolution of research paradigm   154, 158
Explanatory case study   98–100, 110
Explanatory design   8, 87, 88, 91, 93, 94
Exploratory case study   98, 99,
Exploratory design   8, 87, 88, 91, 92, 94
External factors   122, 124
Extreme-condition test   46
Eye-tracking   121, 122, 124, 125

f

Focused group   12, 24, 25, 106, 173–175
F-test   42

g

Generalisation   136
Generalisation of fresh theoretical positions   200
Grand theory   10
Grounded theory   5, 21, 25, 27, 29, 31–35, 92, 93

h

Heterogeneous population   30
Highlights   180
Histcite   172
Hypothesis   2, 3, 7, 9, 10, 11, 32, 37–39, 41–43, 80, 95, 98, 100, 104, 108, 109, 114, 121, 153, 191, 198

i

Independent variable   38–40, 100–104, 109
Induction   5, 11, 12, 16, 32, 34,
Industry-lead mode   199
Industry-university partnership   199
Information and communication technologies (ICT)   151, 156
Innovation   189, 191–193, 196–198
Instrumental case study   98–101
Integral error test   47
Interpretative   20, 85, 97,

Interpretive structural modelling   70, 78–80
Interrelationship   1, 10, 40, 71, 73, 152, 169, 172
Interventional factor   122, 124
Intrinsic case study   98–101

l

Latent Dirichlet Allocation (LDA)   33, 34
Logical inference   5, 16, 27
Longitudinal case study   98–100
Longitudinal research framework   226–232

m

Mediating variable   38, 39, 40
Middle-range theory   10
Mixed methods research   7, 8, 14, 19, 21, 80, 85–87, 89, 90–92, 94, 95, 113, 131, 153, 155, 163, 166, 216, 218
Moderating variable   38, 39,
Modification   200, 212

n

NDLTD   197
Nvivo   32, 164, 172, 175

o

OATD   197
Observability   10
Ontology   4, 16, 85, 157, 218

p

Paradigm   1, 3, 5, 7, 9, 10, 14–16, 19, 22, 85, 97, 114, 130, 151–158
Philosophy-methodology-method-content-context   218
Physical factor   122, 124, 126
Positivism   3, 4, 5, 85, 218
Positivist   4, 20, 97, 98
Postpositivism   3, 4, 85
Practical problem   190, 192, 207–209, 213
Pragmatism   3, 4, 94, 153
Predictive analytics   136
Prescriptive analytics   136
Proposition   1, 3, 10, 40, 88, 201, 216
ProQuest   197
Psychological factor   122, 124, 126

q

Qualitative comparative analysis
(QCA)   110–114
Qualitative Research   5, 7, 10, 11, 14, 16,
19–21, 23, 25–27, 29, 32, 34, 35, 61, 80,
85–94, 113, 129, 134, 163, 173
Quality of interview dialogue   29
Quantitative research   4, 5, 7, 9, 10, 14, 15,
20, 25, 37, 40, 80

r

Reflective thinking   220
Research ethics   117, 126, 148, 188, 219, 229
Research scope   27, 30
Research-practice gap   199, 200,
Responses to reviewers' comments   181
Rhetoric   5, 16

s

Sample size   26–30, 34, 41, 89, 111, 155
Scientific research problem   207–209
Self-conscious integration of theory and
practice   201
Semi-structured interview   24, 30, 31
Sentiment analysis   33, 34
Sequential pattern mining   136
Social network analysis   61, 65, 66, 70, 71, 80
Status of submission   181
Stock-flow diagram   45, 46, 48
Structural equation modelling   7, 41–43, 80,
92,
Structure test   46
Structured interview   24, 25, 30, 31
Structure-oriented test   46

Suggesting referee   181
System dynamics   43, 46, 48, 54, 80, 81

t

Target audience   211, 213, 223
Text mining   33
The fifth research paradigm   152, 156–158
The first research paradigm   152, 154, 155,
The fourth research paradigm   154, 157,
158
The second research paradigm   152–155
The third research paradigm   153, 154
Theoretical approach   201
Theory development   200, 212
Time series analysis   136
Title page   180
Training and integration   200, 212
Triangulation design   8, 87, 89, 91, 93, 94
Trivial theory   10
t-test   41

u

Underpinning theory   30
Unstructured interview   24, 30

v

Verifiability   10
VOSviewer   172

w

Web crawler   131, 133, 134,
Worldview   3, 4, 5, 16, 166, 216, 218

z

z-test   41